Safety Management

Despite efforts to reduce workplace accidents through technical innovations and detailed safety procedures, incident rates remain high. Studies have confirmed that the most common causes of workplace accidents are rooted in human factors including rushed decisions, inattention, risky habits, complacency and misused equipment. Organisations may invest heavily in systems and tools, but if the psychological and social aspects of human behaviour are overlooked, these investments often fail to deliver. Understanding how individuals think, feel and interact in group settings is essential to creating a genuinely safe and resilient work environment which this book explores.

This book investigates why technical and organisational solutions alone are not enough to ensure workplace safety. Drawing from psychological theory and real-world experience, it demonstrates how integrating an understanding of human behaviour into safety management systems leads to better and more effective outcomes. Through detailed case studies from major international enterprises operating, readers will see how practical, low-cost programmes can be used to shape safer behaviours and attitudes at every level of an organisation. The reader will develop a deeper, more sustainable view of what a good safety culture looks like that can align human nature with workplace expectations.

Safety Management: Psychological Aspects of Implementing Organisational Change is for a wide range of professionals involved in shaping workplace safety, including managers, team leaders, OHS specialists, HR professionals, compliance officers, consultants and operational staff in high-risk industries.

Safety Management

Psychological Aspects of Implementing Organizational Change

Katarzyna Boczkowska, Marta Stasiła-Sieradzka, Marta Znajmiecka and Olga Zwardoń-Kuchciak

CRC Press
Taylor & Francis Group
Boca Raton London New York

CRC Press is an imprint of the
Taylor & Francis Group, an **informa** business

Designed cover image: iStock

First edition published 2026
by CRC Press
2385 NW Executive Center Drive, Suite 320, Boca Raton FL 33431

and by CRC Press
4 Park Square, Milton Park, Abingdon, Oxon, OX14 4RN

CRC Press is an imprint of Taylor & Francis Group, LLC

First edition published by CRC Press 2026

ISBN: 978-1-041-15099-2 (hbk)
ISBN: 978-1-041-14383-3 (pbk)
ISBN: 978-1-003-67774-1 (ebk)

DOI: 10.1201/9781003677741

Typeset in Times
by SPi Technologies India Pvt Ltd (Straive)

Contents

About the Authors

Katarzyna Boczkowska is a Doctor of Economics in the discipline of management science. Currently working as an Assistant Professor at the Institute of Marketing and Sustainable Development at the Technical University of Łódź. For more than 20 years, she has been the head of postgraduate studies in 'Occupational Health and Safety Management'. She has 15 years of experience as an expert witness, having drawn up more than 200 opinions commissioned by courts and prosecutors' offices in Poland. She has experience as a trainer and has carried out implementation projects on OHSAS 18001, ISO 45001 management systems, and Behavior-Based Safety programs. Her interests include occupational health and safety management in a broad sense, in particular: occupational risk management, safety culture, BBS, modeling of accidents at work, and OSH management systems. She is the author of more than 100 scientific publications, including monographs and articles.

Marta Stasiła-Sieradzka is a Doctor of Humanities, work and organizational psychologist, research and didactic employee at the Institute of Psychology, University of Silesia in Katowice, Poland. Author of scientific projects and implementations related to numerous initiatives in the field of management and shaping work safety culture for companies from the metallurgical, mining, construction, refinery, and many other sectors. As part of her teaching and training activities, she conducts thematic classes on change management, motivational systems, and creating safe behavior in organizations. She is an adjudicator in the field of psychology of road transport, rail transport, and mining work. Author of numerous scientific publications in national and international journals and a popularizer of science. In 2023, she won the Third Degree Team Prize in the National Competition for the Improvement of Working Conditions organized by the Minister of Family and Social Policy and the Central Institute for Labor Protection in the category of scientific and research works for the study and creation of safety culture in organizations.

Marta Znajmiecka is a Doctor of Humanities in Psychology, work and organizational psychologist, research and didactic employee at the Institute of Psychology, University of Łódź, Poland. Author of scientific projects and implementations related to numerous initiatives in the field of management and shaping work safety culture for, among others, companies from the food, construction, and refinery sectors. As part of her teaching and training activities, she conducts thematic classes on safety culture, safety psychology, and prevention. Creator of a questionnaire to measure safety climate and attitudes towards safety. Author of numerous scientific publications in national and international journals, and popularizer of science. In 2023, she was awarded the Third Degree Team Prize in the National Competition for the Improvement of Working Conditions organized by the Minister of Family and Social Policy and the Central Institute for Labor Protection in the category of scientific and research works for the study and creation of safety culture in organizations.

Olga Zwardoń-Kuchciak is a Psychologist, educator, and business and interpersonal skills trainer and research and didactic worker at the Institute of Psychology, University of Łódź, Poland,in the Department of Work Psychology, Organization, and Career Counselling. In her research work, she explores the development of psychological capital and work crafting in the context of employees' professional effectiveness and well-being. She is a member of the international GIG-OSH project team and a contractor in the RESET project, implementing training activities in the field of diversity & inclusion. In her coaching work, she specializes in the development of psychosocial skills related to emotion and tension regulation, communication, relationship building, and well-being in the workplace and beyond. She is a speaker at webinars, scientific, educational, and business conferences. She designs training programs and scenarios in both traditional and remote forms (e-learning, remote training).

Preface

Security in organisations is not just a matter of technology, procedures or management systems; it is first and foremost about people – their decisions, behaviour and attitudes. While they are the most important element of any structure, they are also its weakest link. Even the most advanced security systems can fail if employees do not follow the rules, underestimate risks or act under misconceptions. In an effort to minimise risks, organisations introduce norms, standards and controls, often forgetting a fundamental aspect – human nature. It is this that determines whether procedures will be followed or remain only on paper.

The book *Safety Management: Psychological Aspects of Implementing Organisational Change* addresses a key challenge of contemporary safety management: how to effectively combine formal structures with the psychological mechanisms of human behaviour. How can we encourage employees not only to follow the rules, but also to feel a shared responsibility for safety? How can we implement change in an organisation to avoid resistance and instead build commitment?

The first part of the book examines safety culture – a phenomenon that has been of interest to both theorists and practitioners for years. Although the concept gained prominence after the Chernobyl disaster, its roots can be traced back to the early 20th century, when it was noticed that employees in different organisations develop their own norms and values. Today, we know that safety culture is not just a set of procedures, but also a way of thinking – a perception of risk and mutual responsibility. In Chapter 1, we look at its definitions, models and methods of measurement, showing how it can be effectively diagnosed and shaped.

Chapter 2 focuses on the concept of Behaviour-Based Safety (BBS). It is an approach that goes beyond traditional training and inspections, focusing on observation, analysis and reinforcement of positive habits. BBS is not a one-off action, but an ongoing process that requires the involvement of all employees – from management to the rank and file. We present both the theoretical underpinnings of the method and the practical challenges of implementing it.

Change is an intrinsic part of modern organisations. In Chapter 3, we focus on the psychological aspects of change management, showing how it is crucial to consider human reactions – from resistance to motivation. Successful transformation requires not only new procedures, but also understanding, communication and support. Without these, even the best strategies can fail.

The final chapter introduces the concept of job crafting – the active shaping of work by workers themselves. It is an approach that allows them to modify tasks, relationships and perceptions of their role, resulting in greater commitment and a sense of responsibility for safety. Job crafting is not only a tool to reinforce a culture of safety, but also a way to adapt to changes that are becoming increasingly dynamic and unpredictable.

This book is aimed at both academics and practitioners – managers, safety professionals, trainers and anyone who wants to understand how psychological aspects affect the effectiveness of safety management. Our aim is to show that true safety does not depend only on systems, but on people – their attitudes, motivation and readiness to act. Only by combining formal structures with psychological understanding is it possible to build an organisation in which safety becomes a natural part of the culture and not just an obligation.

The unquestionable value of the publication is the examples of practical activities presented, which can be used as material for both exercises and training. They can also provide guidance for practical activities carried out in an organisation. What is worth emphasising is that these exercises and case studies not only enrich the learning process and action planning but can also function as a key element of benchmarking – enabling organisations to systematically compare their own safety management practices with effective models based on a psychological understanding of people. This creates an opportunity for continuous improvement and strengthening of safety culture.

We invite you to read what we hope will inspire reflection and action for safer workplaces!

1 The Current Understanding of Safety Culture and the Tools for Shaping It

Over the years, we have seen significant changes in the level of safety in plants. We have gone from very unfavourable accident and injury statistics at the turn of the 20th century to a significant reduction. The positive impact was made possible by the development and implementation of a number of safety factors. In the initial phase, the reduction in accidents and incidents was linked to improvements in technical equipment, involving, among other things, the acquisition and use of knowledge about the properties of the materials used, the use of safer solutions and safety systems. In the next phase, the focus was on the person doing the work. Then it was on improving the performance of employees through appropriate selection and screening, an extensive training system and the introduction of appraisal and motivation systems. This was followed by the introduction of appropriate management systems which use the results of hazard analyses and risk assessments to optimise decision-making processes. In recent years, a high level of safety has increasingly been linked to safety culture (Markowski, 2022).

Although the popularity of the term *safety culture* was undoubtedly fostered by the International Atomic Energy Agency's report produced after the Chernobyl disaster (IAEA, 1991), it is worth remembering that the first mentions of the role of safety culture in a company began to appear in the literature as early as the beginning of the 20th century, when it was noticed that employees in one organisation develop their own distinct norms, values and ways of behaviour. A pioneer in recognising the signs of a safety culture was De Blois (Casey et al., 2022). He described the phenomenon of a sense of value and priority for safety that pervaded the industrial plant and called it a "safety atmosphere". He also signalled the role of the process of learning by imitation, emphasising that observing the behaviour of experienced colleagues has a greater impact on workers' safety behaviour than training.

The term *safety culture* has permanently entered the vocabulary of specialists, theorists and practitioners dealing with security issues. The phenomenon is of great interest. It has generated a significant amount of academic research and activity undertaken by practitioners (Bye et al., 2020). Van Nunen, Li, Reniers and Ponnet (2018) found 1,789 publications on safety culture in the Web of Science between 1900 and 2015. Described by some as a remedy for the problems of incremental change in safety management (Hudson, 2007) and dismissed by others (Hopkins, 2019), safety culture has generated not only academic debate, but also considerable

DOI: 10.1201/9781003677741-1

confusion among practitioners (Le Coze, 2019). Moreover, research on safety culture has spawned a plethora of versions (dimensions) of it, such as patient safety culture (Fleming, 2005), traffic safety culture (Edwards et al., 2014), process safety culture (Olive et al., 2006), nuclear safety culture (Lee & Harrison, 2000), resilient safety culture (Adjekum & Terns, 2020), strong vs. weak safety culture and even "emotional" safety culture (Wang et al., 2018).

Rather than a broad literature review, which is beyond the scope of this chapter, we focus on the key issues related to safety culture. We concentrate on its conceptualisation, selected research findings and the problems of diagnosing this phenomenon. We also demonstrate, using examples, how safety culture can be effectively diagnosed. We believe that the content presented will also be helpful in discussions with employers, providing factual arguments for investing in a safe and friendly working environment.

1.1 SAFETY CULTURE AND SAFETY CLIMATE AS AN ELEMENT OF ORGANISATIONAL CULTURE

The relationship between safety and organisational culture can be well understood by referring to one of the best known concepts: Edgar Schein's (2010) organisational culture model. In it, the author distinguishes three levels of culture, which differ primarily in their degree of awareness and visibility. The deepest and least visible is the third level, which includes the basic, often unconscious, organisational assumptions and beliefs. These largely shape the priorities for action, including in the area of security. At level two are the values and norms supported by the organisation, which express what is considered important and desirable. Level one comprises the most visible elements of culture – the so-called "artefacts" (i.e. the symbols, language, behaviours or rituals present in the day-to-day functioning of the organisation). The hidden assumptions (level three) are formed over time in response to challenges of internal integration (e.g. cooperation between employees) and external adaptation (e.g. responding to threats from the environment, including competition).

The concept of organisational culture in relation to safety issues is well illustrated by the model proposed by Geert Hofstede (2000), who compares culture to the layers of an onion. Each layer reflects a different level of depth and permanence of the elements that make up organisational culture, ranging from the most superficial and easily changed to those that are deeply rooted and difficult to grasp and change.

The outer layer consists of symbols – words, gestures, clothing or objects – that are understood only by members of a particular organisation. In the context of safety, these could include, for example, the markings of danger zones, the colours of helmets or specific language relating to protective procedures.

Heroes (i.e. individuals considered as role models) represent the second layer. Their approach to risk and compliance with safety rules reveal what values are genuinely esteemed in the organisation. If an employee who is willing to break procedures in the name of achieving goals is regarded as a hero, this may indicate that safety is only formally acknowledged and that the true value is efficiency or profit.

Rituals (i.e. repetitive actions with social significance) constitute another layer. In the area of safety, these could include mandatory health and safety training, daily briefings or the manner in which protective clothing is worn. In organisations with a strong safety culture, these rituals are observed with great care; for example, sleeves are always buttoned, and safety shoes are worn according to the rules.

These three layers – symbols, heroes and rituals – make up what are known as practices (i.e. elements of culture that can be seen from the outside). Although they are visible, their proper meaning is understood mainly by those operating within a given organisational culture.

At the heart of the model are values – the deepest, most enduring element of culture. These are core beliefs that are formed in the early stages of an organisation and often remain unconscious. In the context of safety, the value placed on the protection of health and life manifests itself in the form of an unwritten code of conduct. It may be reflected in formal bylaws, procedures or standards of behaviour in emergency situations, but also, more importantly, in the day-to-day choices and priorities of organisational members.

A high safety culture is not just based on the existence of rules and procedures, but on a deep-rooted belief that safety is a real value and not just a formal requirement.

According to Kennedy and Kirwan (1998), safety culture, as an element of the overall organisational culture, is a combination of individual and group thought processes, feelings and behaviours, which – in turn – evoke a particular way of looking at an organisation's various safety problems related to the specifics of its operation. Studenski (2000) emphasises that an organisation's safety culture is made up of employees' attitudes towards safety, generally accepted norms of behaviour in this area, as well as the values attributed to health and life. An organisation's safety culture refers to the state of awareness of hazards, defined norms of behaviour in situations of their occurrence and technical and organisational ways of ensuring employee safety (e.g. constant readiness of emergency services, restrictive monitoring of the technical performance of hazard alarm devices, formulation and training in the implementation of specific procedures for dealing with hazards). Safety is a process that requires both environmental adaptation and internal cooperation and coordination. The characteristics of this process fit well with the potential solutions enabled by organisational culture – for example, by defining how hazards should be managed, what constitutes unacceptable or acceptable risk and how unsafe behaviour should be dealt with – in other words, how safety should be understood and treated by members of the organisation (Casey et al., 2022).

1.1.1 Safety Climate as Part of Organisational Climate

Sometimes the terms *safety climate* and *safety culture* are treated interchangeably. Climate is taken as an indicator of culture, and its monitoring is seen as a measure of the effectiveness of the organisation's efforts in shaping safety culture. Although there are many similarities between culture and safety climate, treating these concepts identically is a mistake. The existence of two levels of safety culture was pointed out by Cooper (2000). The first is visible and related to the organisation's safety management itself. Examining it involves auditing the procedures used and the workplace in terms of safety, reviewing the causes and consequences of accidents that have occurred and assessing the remedial mechanisms put in place. The second, implicit level relates to psychological factors, behavioural patterns, values and attitudes, as well as norms adopted in the organisation, along with their perception by employees. It should be stressed that the study of this dimension involves employees and cannot be carried out by researchers simply by analysing documentation. This implicit dimension is referred to as the safety climate of a given company (Znajmiecka-Sikora, 2014).

Initially, safety climate was defined as the sum of perceptions shared by employees about their work environment (Zohar, 1980). Choudhry, Fang and Mohamed (2007) emphasised that safety climate is a reflection of employee perceptions of the organisational management system, including policies, practices and procedures that indicate how safety is implemented in the work environment, treating it as a "snapshot" that describes "how we do certain things here and now". Currently, researchers define safety climate as a dimension of an organisation's safety culture, its hidden layer defined by employees' attitudes towards various aspects of safety in the organisation (Stasiła-Sieradzka et al., 2020). In the discourse on the concept of safety climate, it can be seen that the unifying element of the conceptual apparatus is the role of employees and their perceptions of various aspects of safety in the organisation, which translate into the attitudes and behaviours they present towards safety (Cooper, 2000; Dedobbeleer and Béland, 1991). The safety climate is not just a temporary "snapshot" of safety culture, but a reflection of the layer of "values" held within the organisation. It represents the intersection between core assumptions, stated values and actual practices (the outer layer), thus informing how important safety is compared to other organisational goals (Casey et al. 2022).

1.2 SAFETY CULTURE AS PART OF NATIONAL CULTURE

When analysing the phenomenon of safety culture, many researchers take a broader perspective, treating safety culture as an element rooted in national culture rather than just a separate organisational phenomenon. The work of Hofstede (2000), a Dutch engineer and social psychologist, is fundamental in considering the importance of cross-cultural differences, including those relating to safe behaviour in the work environment. In his classic work *Culture's Consequences*, he presented the results of many years of research aimed at answering the question: what influences people's behaviour more – the culture of the company in which they work or the culture in which they were brought up. In the 1970s, he analysed more than 100,000 questionnaires from employees of the multinational corporation IBM, specifically middle managers coming from 80 countries. The results showed that the culture of origin determines how people react. The data obtained also made it possible to distinguish so-called "cultural dimensions", facilitating the classification of nations in relation to different values. The terms used by the researcher are, therefore, *dimension* and *national culture*. A dimension is an aspect of a phenomenon that can be measured and quantified, while national culture is the collective programming of the mind as a result of growing up in a particular country (Hofstede, 2000). Hofstede's model of national culture assumes that the citizens of a country, through education, political and economic systems, religion and media, develop a common set of core values and practices that influence and normalise the way people behave and think. It is assumed that these values carry over into organisational life through specific behaviours, norms for relating to authority figures and how individual or collective achievements are rewarded, implicitly shaping organisational culture and behaviour, including safety culture. The researcher, continuing his work, distinguished six dimensions of national culture (Hofstede, 2000), each of which can be described on a continuum as low or high. Table 1.1 presents the characteristics of the individual dimensions of national culture, together with an indication of their possible effects on safety culture.

TABLE 1.1
Dimensions of National Culture and Safety Culture

Dimension Characteristics	Examples of Possible Effects on Safety Culture
Uncertainty avoidance	
• Refers to the extent to which people in a given society try to minimise the anxiety caused by new, risky and ambiguous situations. • The feeling of uncertainty is expressed by strong stress and the need for predictability, which is satisfied by various types of laws, regulations and customs.	• New and ambiguous situations resulting from the appearance of unknown or poorly recognised hazards in the organisation can result in high levels of stress, which reduces the feeling of safety among employees and increases the risk of accidental incidents. The organisation can deal with this by introducing various types of rules and procedures. • Countries with a high intensity of the uncertainty avoidance dimension (e.g. Japan) favour formal safety procedures and strict adherence to rules, thus fostering a safety culture based on "procedural safety". • Countries with a low intensity of the uncertainty avoidance dimension (e.g. the US) prefer flexibility, modifying procedures as needed, which may increase risk, but also allows for a faster response in the event of an emergency and fosters safety innovation.
Long-term orientation	
• This indicator indicates the time horizon in a given society and the importance given to the future or the present and the past.	• Long-term orientation can manifest itself in prevention, strategic safety management and investment in modern machinery and modern personal protective equipment.
Indulgence	
• Determines the degree to which society allows individuals to indulge their urges and impulses. A high index indicates an acceptance of natural impulses, an agreement to experience/feel pleasure and a low level of discipline and restraint.	• A high index of permissiveness can mean acceptance of risk, less rigorous norms, consent to breaking safety standards and low levels of discipline.

(Continued)

TABLE 1.1
Dimensions of National Culture and Safety Culture

Dimension Characteristics	Examples of Possible Effects on Safety Culture
Masculinity–Femininity	
• A high masculinity index characterises cultures with a high need for power, ambition, high competition, materialism and assertiveness; gender role differences are very high. • A high femininity index characterises cultures based on cooperation, concern for others and quality of life.	• A high rate of masculinity is associated with prioritising productive goals over security (a "get the job done" mentality); engaging in risky behaviour can be seen as a sign of strength and courage. • In cultures with high rates of femininity, there is a greater emphasis on employee well-being and cooperation.
Power distance	
• This indicator illustrates attitudes to the phenomenon of inequality between people in different countries. • Communication is usually one-way: from top to bottom. • A low indicator is characteristic of democratic countries, open to public consultation. A high one characterises countries with authoritarian governments.	• It affects the communication of threats. • High power distance results in fear of punishment, which consequently contributes to less frequent questioning of decisions by superiors, even when they threaten security, and to a failure to communicate the threat. • In countries with low power distance (e.g. Sweden, Norway), safety communication is characterised by a high degree of openness – employees are actively involved in identifying risks.
Individualism vs. collectivism	
• A high rate of individualism characterises societies where ties between individuals are loose. The individual is primarily focused on themselves and their immediate family.	• Decides whether a mistake is the fault of the individual or the system. • High individualism is associated with the perception of the employee as an "economic unit" to carry out specific tasks, with a lack of interest in the employee/co-worker. • In individualistic countries (e.g. the US, Europe), there is an emphasis on personal responsibility ("blaming the worker"), which inhibits risk reporting. • In collectivist countries (e.g. Japan), safety is treated as a team commitment (systemic error analysis, with no blaming of individuals).

Source: Own compilation based on Hofstede, G. (2000). *Cultures and organisations: Programming the mind*. PWN.

The issue of the relationship between national culture and safety culture gains particular importance in the context of conducting global business and the need to unify management systems. This unification should enable the effective implementation of these systems in diverse cultural circles. Ignoring deeply held cultural values when designing and implementing safety management systems can lead to failures, especially in multinational corporations where cultural differences influence risk perception and organisational behaviour.

Previous research on the influence of national culture on the formation of safe behaviour in the workplace, mainly based on the Hofstede model, has focused on sectors such as construction, shipping, energy and aviation (Mark et al., 2015). The results of these studies confirm that cultural values such as uncertainty avoidance, power distance and long-term orientation significantly influence risk perception, adherence to procedures and willingness to report risks.

1.3 PRACTICAL MODELS OF SAFETY CULTURE

In the 1990s, the concept of safety culture underwent a significant transformation, evolving from theoretical models towards practical applications. By defining safety culture as "a set of characteristics, attitudes and behaviours of individuals, organisations and institutions that ensure that safety is given the highest priority and receives attention commensurate with its importance" (IAEA, 1991), the International Atomic Energy Agency laid the foundations for a practical approach to the concept. This broad definition has made it possible to integrate a variety of concepts into the field of safety culture and has contributed to its effective implementation in various organisations.

The practical approach is based on safety culture maturity models, which consist of multiple dimensions and indicators organised according to levels of effectiveness or stages of development (Goncalves Filho & Wotcrson, 2018). One of the most recognised models is Hudson's five-stage safety culture maturity ladder.

According to Hudson's (2007) model, as organisations progress through successive levels of maturity (from pathological to reactive to calculative to proactive to generative), there is an increase in social capital (mutual trust, collaboration and relationship networks) and an improvement in the quality and transparency of safety information flows. This, in turn, is expected to potentially lead to improved safety performance and greater proactivity and initiative on the part of employees. It should be emphasised that the use of the term *automatically* in describing this relationship is a simplification, as the actual improvement of safety performance is a complex process, dependent on many other factors such as the adequacy of resources, the effectiveness of management systems or the commitment of leaders (Flin et al., 2000). An increase in trust and information flow creates a foundation and favourable conditions, but it is not a deterministic mechanism.

Despite these caveats, it is worth highlighting that, when applied carefully, safety culture maturity models can be valuable diagnostic and strategic tools (Goncalves Filho & Wotcrson, 2018). Their practical applications include (1) strategic security planning, allowing realistic goals to be set; (2) designing interventions that direct efforts towards the most critical areas of development; and (3) measuring and monitoring progress in shaping a safety culture, which allows for the assessment of the effectiveness of actions taken. Moreover, these models are flexible enough to be adapted to reflect the nuances of local organisational contexts, such as industry specifics, legal regulations or prevailing working practices. Safety maturity models, especially in the context of critical analysis, can play an important role in deepening reflection on the concept of safety culture. They contribute to a better understanding of the various factors shaping organisational safety practices and make it possible to distinguish approaches based on sound theoretical and empirical foundations from those of a superficial or unverified nature (Le Coze, 2019).

1.4 THE IMPORTANCE OF SAFETY CULTURE IN CREATING A HEALTHY AND SAFE WORKING ENVIRONMENT

Despite the diversity of research approaches and methods, which lead to results that are ambiguous and difficult to replicate, it is worth taking a closer look at published reports on safety culture. They make it possible to define areas of interaction and relationships between safety culture/climate, work environment factors and worker well-being. Properly identifying these areas enables an understanding of the function of safety culture in an organisation, which is crucial for effective risk management and improved working conditions. This provides a better understanding of how safety culture affects employee safety and health, as well as its potential benefits to the organisation.

1.4.1 Safety Culture and Accident Rates

Analysis of the relationship between safety culture and workplace accidents is based on the assumption that companies with a good safety climate have fewer accidents and injuries compared to those with a poor one. Although the published research reports in this area are not conclusive, it should be emphasised that they primarily indicate a relationship between safety climate, accidents, near misses and the number of risk behaviours undertaken. According to ISO 45001: (1) incident – occurrence arising out of, or in the course of, work that could or does result in injury and ill health; (2) an incident where injury and ill health occurs is referred to as an "accident"; (3) an incident, where no injury and ill health occurs, but has the potential to do so, may be referred to as a "near miss", "near hit" or "close call". Positive relationships between safety climate and accidents at work have been identified in the UK among chemical and marine workers (Donald & Canter, 1994; Mearns et al., 2003); in Hong Kong, California, and the US, among construction workers (Gillen et al., 2002; Siu et al., 2004); in Finland, among woodworkers (Varonen & Mattila, 2000); and in Poland and Israel, among those employed in the metalwork industry (Milczarek & Najmiec, 2004).

1.4.2 Workers' Health and Safety Culture

The definition of health, contained in the 1946 preamble to the constitution of the World Health Organization (WHO), treats it as a state and identifies it with complete physical, mental and social well-being – not just the absence of disease symptoms (WHO, 1946). Representatives of the social sciences promote consideration of its dynamics. Health is defined as a disposition that enables an individual to function adaptively in a specific environment or as a process of seeking and maintaining a balance between the demands of the external and internal environments and the efforts made to adapt.

Published research findings indicate that safety climate is an important determinant of worker health (Bergheim et al., 2015; Schneider et al., 2017; Thakre & Shroff, 2016). According to Golubovich et al. (2014), a low level of organisational safety climate is a stressor that causes emotional tension among employees. It is associated with the emergence of physiological (e.g. increased muscle tension), behavioural (e.g. risk-taking) and psychosomatic (e.g. lower pain threshold) symptoms, which explains the development of musculoskeletal disorders.

Tamakloe et al. (2022) investigated the role of fatigue as a mediating factor in the relationship between safety climate and worker well-being in the aviation sector, a high-risk area requiring strict adherence to safety standards. The study included 305 operational workers at Kotoka International Airport. A mediation analysis showed that fatigue partially mediates this relationship, meaning that safety climate influences employee well-being both directly and indirectly through fatigue reduction. Creating a work environment perceived as safe can, therefore, reduce fatigue and improve workers' psychological well-being.

Zadow et al. (2021), based on a study of 1,084 Australian full-time workers, indicated that a high level of safety climate can prevent symptoms of depression in workers.

1.5 THE DETERMINANTS OF SAFETY CULTURE – A REVIEW OF THE RESEARCH

In parallel to research aimed at clarifying the role of safety culture in shaping a safe and healthy work environment and in promoting employee well-being, efforts have been made to identify the factors that contribute to organisations achieving a high level of safety culture. The analysis considered the importance of organisational variables, including organisational climate, leadership style and safety management system. In an effort to identify universal determinants, comparisons were made across industries and countries.

1.5.1 Organisational Culture and Climate versus Safety Culture

There is no doubt that organisational culture can have a significant impact on workplace safety. Not only can flaws in the organisation of the work process but also profit-maximising policies encourage or even lead to risky behaviour, resulting in accidents or loss of health. Research by Neal and Griffin (2006) showed that organisational climate has a significant impact on safety climate and safety-related employee behaviour. The results of the study confirmed the role of safety climate as a predictor of safe behaviour in the workplace and as a mediator of the impact of organisational climate on safety-related outcomes. Similar results were obtained in a study by Colley (2013), who analysed the relationship between organisational values and workplace safety. The researchers distinguished between four organisational profiles that differed in their perceptions of organisational values, psychological safety climate and safety incidents. The results of the study showed that those who perceive their organisation as emphasising employee well-being rate the safety climate higher and report fewer safety incidents.

Auzoult and Gangloff (2018) examined the relationship between organisational culture and safety climate among 180 employees from the manufacturing, retail, service, health, construction, public administration, security, agriculture and transport sectors. The analyses revealed a positive relationship between organisational rules and safety climate. Cultures with high levels of control (authority, rationality and procedures) or low levels of control (participation, collaboration and open communication) are conducive to the development of safety, as long as they are based on the internal orientation of the organisation, where cohesion and integration are prioritised. The authors suggest that safety is easier to develop in organisations that are internally oriented (focused on employee welfare and development) than in those that are externally oriented (focused on company expansion), advocating further research for the latter.

1.5.2 Safety Culture and the Safety Management System

Systemic safety management is based on the implementation of a standard in the area of health and safety. The current standard is ISO 45001:2018, which replaced earlier ones (e.g. ISO 9001, PN-N 18001, OHSAS 18001). The issue of the relationship between the safety management system and safety culture has been the subject of many studies. Mikkelsen and Saksvik (2004) indicated that the introduction of a safety management system in the electrical industry was associated with a reduction in both the number of accidents and the number of accident incidents, and contributed to an increase in safety culture, manifested in safe behaviour and more positive attitudes towards safety among employees. Similar results were obtained in a Polish study by Znajmiecka-Sikora and Boczkowska (2016), which diagnosed the safety climate in ten companies. The study showed statistically significant differences between organisations that had implemented PN-N 18001 or OHSAS 18001 systems and those without these systems. Enterprises with these systems were characterised by a significantly more positive safety climate, manifested in higher risk awareness and greater employee responsibility, increased staff participation in safety issues, better preparation for emergency situations and the prioritisation of safety alongside production goals. By contrast, organisations without the implementation of the aforementioned systems exhibited a significantly lower level of safety climate, characterised by reduced hazard awareness, diminished responsibility for safety, a lack of promotion of safe working practices and the prioritisation of production aspects over safety.

1.5.3 Safety Climate and Leadership

According to Zohar's (2010) findings, leaders play a crucial role in modelling the safety climate through the manifestation of their attitudes, values and routines. The author showed that workers' perception of safety priorities is largely determined by the visibility and commitment of leadership to aspects of health and safety. In particular, transformational leadership – characterised by inspiration, motivation and personal example – significantly supports the formation of positive orientations towards safety.

Barling, Loughlin and Kelloway (2002), on the one hand, based on the results of a survey of restaurant staff in the US, confirmed that transformational leadership enhances safety in the work environment. Zohar and Luria (2005), on the other hand, analysing leaders' behaviour, concluded that leaders' focus on safety issues in their interactions with employees promotes both a positive perception of the safety climate and improved safety indicators.

In a natural experiment, Zohar and Polachek (2014) showed that targeted interventions in the form of weekly feedback from supervisors to staff significantly optimise safety climate and safety performance. It is also worth citing findings from an analysis of safety culture at Taipei Songshan Airport in Taiwan, where safety oversight, a safety-focused work environment and a reward- and sanction-based incentive system were found to be key determinants of improving an organisation's safety culture (Fu & Chan, 2014). Analogous implications regarding the role of effective leadership in creating safety come from research in the petrochemical sector (Zahiri Harsini et al., 2021).

1.5.4 In the Search for Universal Indicators of Safety Culture: The Role of Industry Affiliation, Department and Position Held

The literature consistently emphasises that perceptions of the safety climate in organisations vary by occupational group (Chen et al., 2017; Håvold, 2010; Stasiła-Sieradzka & Dobrowolska, 2016; Tyszkiewicz, 2019; Yari et al., 2019). Comparative studies have shown significant differences in hazard assessment and perceptions of the safety climate between groups such as miners (underground, open-pit, drilling), surface airport personnel, seafarers, polar region workers, medical personnel, lighting and construction workers, mountain rescuers and retail workers (Gabryelewicz & Krupa, 2016; Nowacki et al., 2016; Schwatka et al., 2016). The results of research conducted in Poland from 2017 to 2019 on a sample of 2,995 individuals from the three identified occupational groups (production line workers, retail trade workers and mountain rescuers) confirm that differences in the perception of safety climate mainly relate to the so-called "soft" aspects, related to the competence of managers and employees, modelling and reinforcement of safe behaviour and work atmosphere. By contrast, "hard" aspects, including standards, norms, technical facilities, ergonomics and the organisation's OSH policy, did not differentiate between the analysed groups, which is explained by the universality of legal regulations and international standards in this area (Stasiła-Sieradzka et al., 2020).

Another key factor that influences the perception of safety climate is an individual's position within the organisational structure. Those in higher positions, who have greater influence over shaping the safety culture, tend to rate the safety climate more positively (Petitta et al., 2017). The differences in perception include management commitment, the modelling of safe behaviours, awareness of risks and knowledge of protective procedures. Conversely, executive employees often exhibit lower awareness of potential threats, a lack of familiarity with protection methods and lower ratings regarding the quality of communication and relationships with supervisors (Znajmiecka-Sikora & Boczkowska, 2016).

It should be emphasised that the identified differences in the perception of the security climate, depending on the industry, department or position, have important practical implications. They point to the need to take action not only in the procedural-technical area, but also in the social and psychological area, taking into account the specifics of individual professional groups. This forms the basis for the design and implementation of prevention programmes aimed at strengthening areas in need of intervention (Stasiła-Sieradzka et al., 2020; Znajmiecka et al., 2022).

1.6 MEASURING SAFETY CULTURE BASED ON A TRIANGULATION APPROACH (A CASE STUDY ON THE EXAMPLE OF SELECTED COMPANIES)

In recent years, the diagnosis of safety culture has attracted considerable interest among both theorists and practitioners. Although the data obtained from them indicate the weaknesses and strengths of the management system, the way in which the measurement is carried out often raises questions. Researchers are confronted with the paradox of *declared* vs. *actual* culture. Employees declare support for safety but prioritise efficiency in critical situations. A study by Bae et al. (2021) showed a discrepancy between perception and reality: (1) surveys suggested that 88% of employees rated the safety culture as "good"; (2) observations showed that 45% of employees ignored protocols in high-risk situations (e.g. speed vs. safety); (3) interviews revealed that employees ignored safety rules due to production pressure and fear of penalties for delays; (4) analysis of documents proved that 30% of "minor" incidents (near misses) were not reported by employees, contradicting the declared rules. In the context of the aforementioned reports, it is worth asking how to study safety culture in order to obtain reliable data. Figure 1.1 illustrates the worthwhile steps in designing and implementing safety culture surveys.

When designing a study of safety culture, the question must be answered as to how we understand culture and on which theoretical concept our diagnosis will be based. Research questions and hypotheses should be formulated, resulting in the definition of the areas of research and analysis to be conducted. For example, we may seek an answer to the exploratory question(s) – What is the safety culture in the organisation?; or Are the preventive actions carried out by the organisation effective; that is, does the organisation teach safety to its employees?; or How does the organisational culture affect the safety culture of the organisation and employee behaviour, attitudes towards safety, individual well-being, and so on? This is of particular importance, as all the methods and measurement tools we will use should allow us to verify the questions and hypotheses posed. At the stage of designing the research, it also seems reasonable to divide it into individual groups according to which the results will be analysed, bearing in mind that the perception of safety climate depends, among other things, on the profession, the position occupied in the organisation or the area of employment (e.g. production worker vs. office worker).

Measuring the phenomenon of safety culture is a very complex process. Culture is like an onion – it has layers. Underneath the top layer (visible to the naked eye) are further layers (hidden), but very important: values, norms and assumptions. It is necessary to access them skilfully and to be able to distinguish between declarative values and those that actually guide behaviour. The level of consistency of the message must also be determined. This requires a comprehensive approach to the research process, combining quantitative and qualitative measurement methods (*triangulation*), which makes it possible to verify information from a variety of sources. This approach enables better access to data and allows the analysed phenomenon to be viewed from multiple perspectives (Stanisławski, 2017).

Quantitative measurement of safety culture is based on the results of surveys and questionnaires. These tools should be checked for psychometric properties. This requires

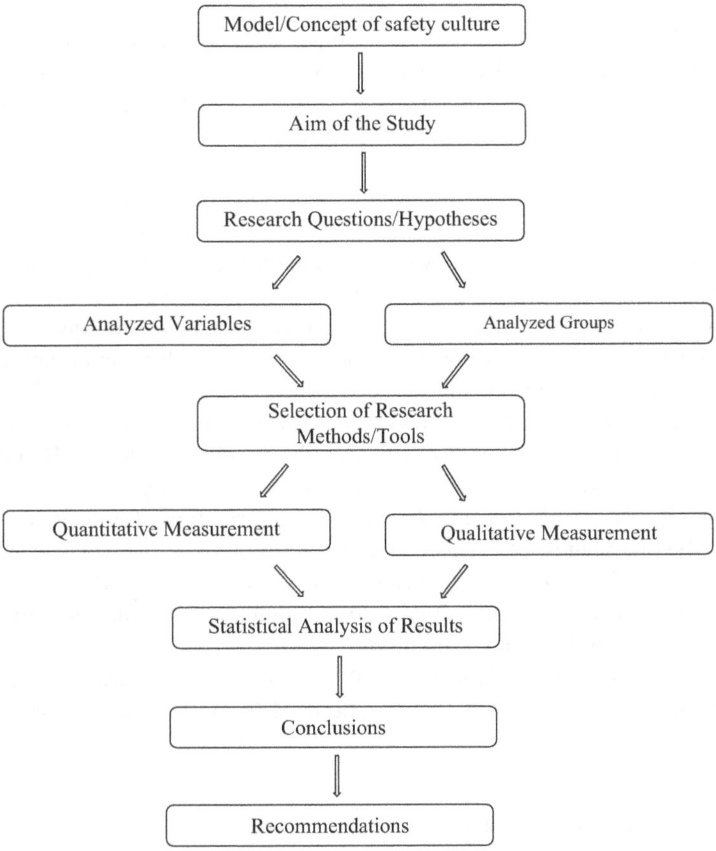

FIGURE 1.1 Measuring safety culture.

Source: Own elaboration.

the use of an appropriate procedure for constructing the tool, taking into account, among other things, the proper (from the methodological side) preparation of the questionnaire, pilot studies and appropriate statistical analyses. This is a complex and time-consuming process, so it seems easier to use existing ones, for example:

- *The Safety Climate Questionnaire (KKB-50)* makes it possible to diagnose the safety climate in ten dimensions: employee participation in safety matters, safe behaviour, management commitment, modelling and reinforcing safe behaviour in the organisation, managing the risk of accidents in the workplace, technical facilities and ergonomics, work pace and fatigue levels, the OSH training process, workplace atmosphere and the organisation's OSH management policy. It allows for the preparation of a map of the strengths and weaknesses of the safety climate of an organisation (Znajmiecka-Sikora, 2019; Stasiła-Sieradzka et al., 2020).

- *Questionnaire on Attitudes to Safety (KPwB).* Using it, we will obtain knowledge about the attitudes that employees present towards the occurring physical, material and psychosocial risks in the work environment and a diagnosis of the three components of attitude – affective (emotional), cognitive and behavioural (Znajmiecka-Sikora, 2019).

Qualitative methods for measuring safety culture include focus interviews, audits, observations, documentation analysis and site visits. In the diagnosis of safety culture, focus interviews are used to assess the internalisation of the principles of safe behaviour, including attitudes to hazards, the understanding and interpretation of the motivational range (penalties and rewards) regarding safe and unsafe behaviour, the authority of formal and informal leaders together with their influence on creating a safe working environment, as well as the pathways of non-safety attitudes and behaviours. On the one hand, audits and reviews of documentation and procedures are key tools for verifying the effectiveness of implemented OSH systems and human resource (HR) management practices, with a particular focus on their impact on employees' subjective sense of safety and the formation of a positive safety climate. On the other hand, site visits and direct observation in the workplace allow for the assessment of the physical organisation of the workspace and safety messages, examining the processes of perception and interpretation of signals and their impact on staff behaviour in hazardous situations.

Figure 1.2, based on Schein's (2010) organisational culture model, provides examples of methods and tools that can be used to measure safety culture. When

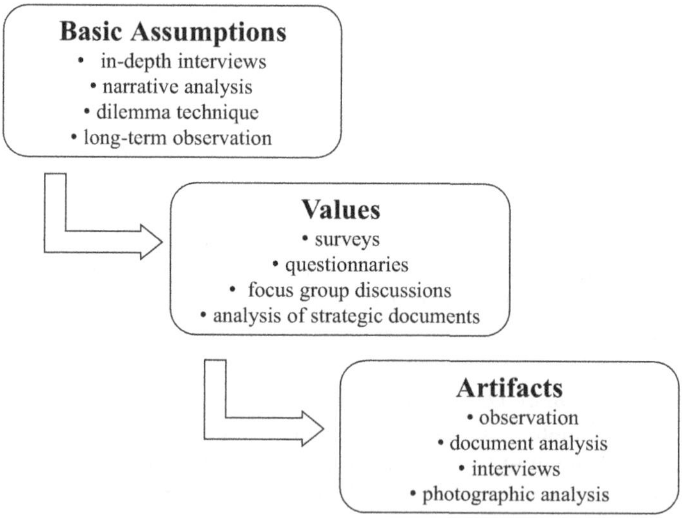

FIGURE 1.2 Examples of methods and tools for measuring safety culture based on Schein's model.

Source: Own elaboration.

deciding which methods and tools to use, it is important to keep in mind both the theoretical model adopted and the purpose of the research (research questions or hypotheses to be verified).

In the research design process, it is also important to ensure measurement on a representative group. It is worth bearing in mind random sampling. It is necessary to ensure voluntary participation and anonymity in the research process. After conducting the research, the data obtained is subjected to statistical analysis, which allows for addressing the questions and hypotheses posed, drawing conclusions and designing further actions – for example, a programme for the development of a safety culture.

CASE 1: EXAMPLE OF A KB FOCUS GROUP STUDY BASED ON SCHEIN'S MODEL

In the focus study, using Schein's (2010) model to diagnose safety culture, we will seek to uncover these three levels, with the biggest challenge and goal being to infer the underlying baseline assumptions from discussing the artefacts and proclaimed values and experiences of the participants.

Issue areas to be discussed:

1. Safety artefacts:
    - Physical work environment (order, signage, availability of PPE and condition of equipment).
    - Procedures and work instructions (their availability, form and degree of compliance in practice).
    - Incident/accident reporting system (what it is like, whether it is used and what happens after reporting).
    - Safety training (its form, frequency, content and how it is conducted).
    - Visual communication (boards, posters, and health and safety slogans).
    - The way meetings are conducted (does it start with H&S?).
    - Language used to talk about safety (e.g. "requirement", "bureaucracy", "our common cause").
    - Visible behaviours of leaders and staff in the context of safety ("walking the talk", responding to breaches and looking after their own safety and that of others).
2. Proclaimed safety values and standards (what the company says):
    - Official safety policy (is it known and understood).
    - Management declarations about the priority of safety.
    - Strategic objectives related to safety (e.g. zero accidents).
    - Reward/penalty systems for safety-related behaviour (what is officially praised/punished).
    - How the company talks about accidents/incidents (e.g. "unacceptable", "lesson for the future", "whose fault is it?").

3. Infiltrating underlying assumptions (through discussion of experiences, contradictions and interpretations):
   - Real safety priority vs. production/cost: What happens when a production objective conflicts with a safety procedure? (Inferring the real priorities of the organisation). Risk perception: Is risk perceived as something to be avoided at all costs or as "part of the job"? (Inferring assumptions about the nature of work and risks).
   - Reporting culture: Do people feel safe reporting problems, mistakes or near misses? Do they fear the consequences? (Inferring assumptions about trust and punishment/blame).
   - Just culture: How is the perceived treatment of people who make a safety-related mistake? Is there a focus on blame or on the system and learning? (Inferring assumptions about human error and responsibility).
   - Learning culture: Is there actually deep analysis and change after incidents/accidents, or just "patching the hole"? (Inferring assumptions about change and learning from mistakes).
   - The role of leaders: Are leaders perceived as genuinely committed to safety, or merely as "demanding"? Do their actions match their words? (Inferring assumptions about leadership and their role).
   - Mutual accountability: Do employees feel responsible for their own and their colleagues' safety? Do they pay attention to each other? (Inferring assumptions about interpersonal relationships and collective responsibility).
   - Trust: Do employees trust that the company cares about their safety? Do they trust the systems and procedures? (Inferring assumptions about employee–organisation relationships).
   - Adequacy of resources: Do employees feel they have sufficient resources (time, people, equipment and training) to do their job safely? (Inferring assumptions about the value of safety in relation to other goals).

Sample focus discussion questions (open-ended, neutral, encouraging stories):

Remember that questions should move from the more concrete (artefacts) to the more abstract (experiences, interpretations) in order to gradually uncover the deeper layers of culture. It is also important to deepen the answers with questions such as: Why do you think so? Can you give an example? What happens in this situation?

Introduction:

The aim of our meeting is to better understand how security is perceived in our organisation. We want to discuss what works well, what could be improved and what *really* influences how we prioritise safety daily. There are no right or wrong answers; all your perspectives are important and will help us create a safer workplace. Our conversation is confidential (explain purpose, rules and confidentiality).

Questions – Level 1: Artefacts (Visible Aspects of Safety)

1. When you look around your workplace (office, hall, construction site, etc.), what comes to your mind first in terms of safety (e.g. signage, order, equipment)?
2. What safety "rituals" do you observe in your daily work (e.g. morning meetings, pre-work inspections, how the shift is handed over)?
3. How are safety issues communicated to you (e.g. whiteboards, emails, apps, direct conversations)? What are these communications like?
4. Do you remember any safety-related stories that are often told at work? What are they about?
5. In your opinion, what are the safety procedures and instructions like? Are they easy to find? Understandable? Do you follow them in practice?
6. What visible behaviours of managers or leaders indicate that safety is important to them (or not)?

Questions – Level 2: Proclaimed Values (What Does the Company Say about Safety?)

7. What does the company say about safety? What are the main messages, slogans and objectives that you hear about from your superiors or in official documents?
8. What do you feel are the stated priorities of the organisation in terms of safety (e.g. safety always comes first, production comes first, it is important that no one gets killed)?

Questions – Level 3: Basic Underlying Assumptions (Experiences, Interpretations and Inferences)

9. Do you have a sense that what the company says about safety is consistent with what happens on a daily basis? Where do you see the biggest differences or the biggest similarities? (Bridge to the assumption of true priorities).
10. Imagine a situation where you see something that is not safe, or you are close to an accident (a so-called "near miss"). What is the first thing that comes to your mind? Would you report it? Why/why not? (Assumptions about reporting culture, trust and consequences).
11. When someone in the team makes a mistake that affects safety, how does the company or supervisor usually react? What is this like in practice? (Assumptions about a culture of fairness, treatment of mistakes and blame).

12. Do you have the sense that after an incident (incident, accident or near-accident) the company actually learns from the mistakes and makes real, effective changes, or does the topic "fade away" after a while? (Assumptions about learning culture and approach to change).
13. When there is time pressure or you must "prove" a production/project result, what usually happens with safety issues? Do you have a sense that certain procedures can be "bypassed" in order to make it in time? What are the unwritten rules then? (Assumptions about real priorities and conflicting objectives).
14. How do you assess the commitment of management/leaders to safety at your level? Are they visible? Do their actions back up what they say? Do you feel their support on safety issues? (Assumptions about leadership, their role and authenticity).
15. Do you feel you have everything you need to work safely (time, appropriate equipment, knowledge/training and support from others)? If not, what is missing? (Assumptions about resources and value of safety in terms of investment).
16. How do teammates approach safety? Do you alert each other to risks? Is there any group pressure regarding safety (positive or negative)? (Assumptions about group norms and collective responsibility).
17. Finally, if you had to describe the "spirit" or "atmosphere" of our company in terms of safety, what words would you use? What *really* matters in this company when it comes to safety? (Trying to get to the heart of the underlying assumptions).

Summary:

We are coming to the end. Is there anything else important about security in our company that we haven't discussed that you would like to add? (Last chance to raise important issues).

Thank you for your participation and valuable perspectives.

Important methodological notes:

Moderator: The role of the moderator is crucial. He or she should be neutral, ask open-ended questions, listen actively, deepen answers, manage group dynamics and not impose their own interpretations.

Group selection: Try to make the group reasonably homogeneous in terms of jobs/work areas (so that they have common experiences), but diverse in terms of seniority or age if possible. Several such groups will be needed.

Analysis: Analysis of the interview transcripts involves looking for recurring patterns, key words/phrases, stories, metaphors and, especially, "contradictions" between what is said (Artefacts, Proclaimed Values) and what is inferred from the participants' experiences and stories (thus inferring the Underlying Basic Assumptions). These contradictions are often the best indicators of deeper, unconscious assumptions.

Confidentiality: Absolute adherence to participant confidentiality and anonymity is key to obtaining candid responses.

Context: Remember that the answers in the focus are the perspectives and feelings of the participants. Cultural diagnosis requires triangulation of data (comparison with results from surveys, audits and incident analysis, for example).

Diagnosis of safety culture conducted on the basis of a triangulation approach allows for the obtaining of reliable and accurate information (according to the research standards used in social and management sciences) about what safety culture looks like and what it is characterised by. It indicates strengths and weaknesses, which can and should be the starting point for designing corrective, preventive or developmental actions. This approach allows for a better understanding of the causes and mechanisms behind employees' behaviours, motivations, concerns or daily practices.

Considering age, gender, length of service and, above all, different jobs and areas of employment in the measurement allows information to be obtained on the homogeneity of the organisation, making it possible to design preventive actions tailored to real needs. Carrying out statistical analyses (e.g. correlation, regression, mediation, moderation analysis) allows answers to be obtained to a number of research questions and hypotheses, indicating relationships between individual variables. The results obtained in this way allow for the preparation of recommendations for further actions, enabling the implementation of the obtained results in such areas as:

- promotion of safety through knowledge transfer,
- building training programmes to enrich the content of regular employee training in OSH,
- supporting development and promotion paths of employees,
- authentication of motivation systems,
- recruitment and selection of employees,
- transfer of knowledge and experience within the organisation,
- communication and risk communication.

And that's not all – the replication of the research after a certain period of time allows the effectiveness of the prevention activities carried out to be assessed. This means the proposed measurement methodology can be used not only for diagnostic purposes, but also to evaluate the process of creating a safety culture.

CASE 2: INVESTIGATING WORKPLACE SAFETY CULTURE IN A PETROCHEMICAL SECTOR ORGANISATION

In order to create a safe working environment, organisations take a number of measures to build safety at the technical, organisational and human levels. They refer to humanitarian values and propose a number of solutions that prove to be insufficient. Despite large-scale prevention efforts, they report the problem – "we have done everything, we have procedures, we have safe machinery and equipment, and yet accidents still happen." However, there are also those who systematically examine the level of safety culture by submitting themselves to external verification by experts. The aim of these activities is to constantly seek the best possible solutions to create a safe working environment, preventing emerging technical, organisational and social risks. It should be stressed that the research process itself strongly activates the workforce, making them the most important carriers of information on potential threats, opening the way for continuous communication within the organisation and uniting the goal of all employees, which is efficient and safe work.

The presented model for investigating workplace safety culture was implemented in a multinational company representing the petrochemical sector. This organisation had high occupational safety standards and was also process-oriented in its preparation for both the research process and the implementation of the resulting solutions. The company declares that caring for the personal health and safety of employees as well as stakeholders is a natural and integral part of its organisational culture and business operations. The safety development policy is based on two parallel pillars: individual personal safety and process safety.

1. Safety culture diagnosis methodology

As a result of the analysis of the literature in the area of theoretical concepts of safety culture, methodologies for its measurement and reports from empirical research, a study was designed based on a triangulation approach (Figure 1.3).

In order to achieve the intended results, the criterion of consistency between the qualitative and quantitative research was applied in terms of group selection, which will enable the development of recommendations for further actions to build a safety culture in the organisation, including but not limited to recommendations for health and safety departments, HR, direct managers and training centres that the organisation has. The focus research complements the qualitative and quantitative results obtained from the questionnaires. The benefits of the methodology used are presented in Figure 1.4.

The project was divided into two phases – research (diagnosis) and development (recommendations).

In the research stage, it was designed to carry out work aimed at gaining knowledge on how safety is managed in the organisation, the level of the organisation's safety culture, the attitudes of the staff towards safety and the personal safety culture of its employees.

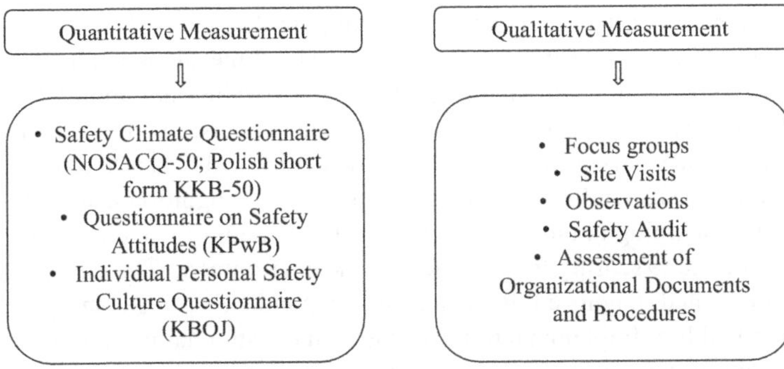

FIGURE 1.3 Methodology for measuring safety culture in a petrochemical sector organisation.

Source: Own elaboration.

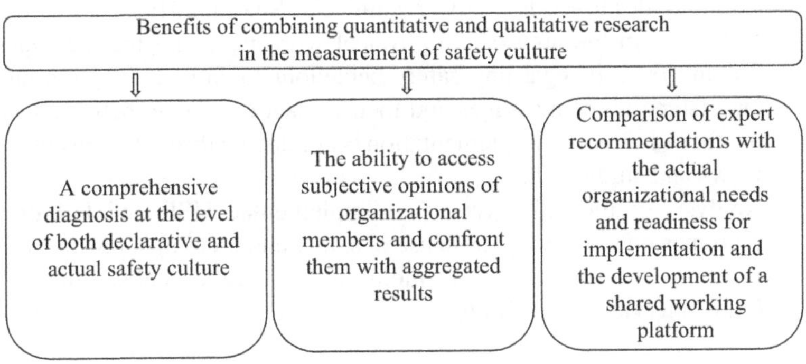

FIGURE 1.4 Benefits of combining quantitative and qualitative research in measuring safety culture.

Source: Own elaboration.

As part of the implementation of the aforementioned phase, the following activities were designed:

1. Identification of the safety climate at work using the Safety Climate Questionnaire (KKB-50) (Znajmiecka-Sikora, 2019).
2. Diagnosis of personal safety culture and determination of its level in relation to risks in the work environment (physical, material and psychosocial) as well as outside the work environment. Implementation was carried out through the creation of the research tool *Questionnaire of Personal Safety Culture of an Individual*, which was prepared for

the needs of the organisation, according to the standards of tool construction used in the social sciences. The dimensions studied were (1) one's own health and its prevention; (2) values and moral dilemmas; (3) ways of protecting oneself from threats; (4) internalisation of social safety principles; (5) attitude to physical risks; (6) attitude to social risks/readiness to help; (7) openness to cognition and exploration, learning about threats; (8) self-assessment of one's own potential; and (9) caring for the safety of the environment. The dimensions presented diagnosed the area of the respondent's private and professional life. Implementation was through questionnaire surveys.

3. Determination of employees' attitudes towards occupational safety, taking into account its affective, cognitive and behavioural dimensions and directional aspect (positive vs. negative attitude) in relation to the physical, material and psychosocial risks present in the work environment. Implementation was through questionnaire surveys. The Questionnaire on Attitudes to Safety (KPwB) was used to measure attitudes towards safety (Znajmiecka-Sikora, 2019).
4. Diagnosis in the area of the physical workspace and the messages flowing from it regarding safety behaviour (area of perception and understanding of messages and their impact on human behaviour in the face of threats). Implementation is conducted through focus interviews and study visits.
5. Analysis of the effectiveness of implemented HSE and HR programmes on improving the sense of safety and creating a positive climate of occupational safety among the workforce, realised through focus interviews and audits.

The area of diagnosis in the process of focus interviews included issues related to:

- the type of risks present in the organisation;
- the extent of employee participation (inclusion) in safety issues;
- employee behaviour – safe vs. risky;
- the level of management commitment to H&S;
- modelling and reinforcing safe behaviour in the organisation;
- how accident risks are managed in the organisation;
- the level of technical facilities and ergonomics;
- the pace of work and level of fatigue experienced;
- the OSH training process;
- workplace climate;
- the organisation's OSH management policy;
- the effectiveness of OSH communication;
- the knowledge transfer process, organisational learning, benchmarking and cultural diversity;

- the process maturity of the organisation;
- the structure of differences between departments in relation to the organisation's safety culture;
- the type of safety objectives set for management;
- the underpinnings of safe behaviour;
- effective forms of supporting safe behaviour.

The *OSH and HR Audit* analysed the effectiveness of the implemented HSE and HR programmes on improving the sense of safety and creating a positive work safety climate among the workforce. Two meetings with each department were planned and conducted. The first meeting, an introductory session, served to gather basic information; the second complemented it. During the meeting with the representatives of the health and safety department, issues relating to, among others, the following were raised:

- employee participation in the creation of a safe working environment and its hygienic conditions,
- organisational learning and transfer of knowledge and experience on safe behaviour,
- knowledge transfer within the organisation on a global level,
- identifying areas of best practice and justifying differences between departments,
- other ways (apart from target setting) to promote global OSH solutions,
- the company image and internal and external communication policy in the context of safety management,
- the prevention-incident-post-accident intervention process,
- the scope of cooperation and communication between HR and H&S departments.

During the meeting with representatives of the HR department, issues were raised concerning the areas of HR activities carried out in the organisation that affect the creation of a positive work safety culture:

- the area of recruitment activities and employee induction in the context of safety,
- training and development in the area of creating safe behaviours,
- promotion and career in the area of safe behaviours,
- motivation systems to enhance safe behaviours in the organisation,
- employee appraisal systems and their parameters related to safe behaviours,
- the extent of cooperation and communication between HR and OHS departments.

Study visits – observations in the area of the physical workspace constituted an observation in the area of the physical workspace and the safety behaviour messages emanating from it. The subjects of the observations were:

- characteristics of the area,
- risks specific to the area,
- type/specificity of work carried out,
- safety messages emanating from the space.

Once the research material was collected, the *Development Stage* was undertaken and a research report was prepared, which included *the development of* assumptions for the programmes: (1) Personal Safety Culture Development Programme; (2) Programme for Building a Positive Safety Climate in the Organisation. It was assumed that the effect of the work would be the development of recommendations enabling the implementation of the results obtained within the framework of the aforementioned programmes in such areas as:

- promotion of safety through knowledge transfer,
- building training programmes to enrich employees' cyclical training in OSH (information on the importance of building a positive work safety climate and the role of personal safety culture),
- supporting staff development and promotion paths (promotion system based on attitude to safety and ability to build a positive safety climate),
- authentication of motivation systems (reward and punishment as elements supporting the building of a positive safety culture at work),
- recruitment and selection of personnel (the level of personal safety culture will be highlighted as one of the selection elements),
- transfer of knowledge and experience within the organisation (solutions will be proposed to support the transfer of knowledge and experience at internal organisational and intergenerational levels),
- communication and risk communication (a communication language will be proposed to support the development of a positive safety climate),
- knowledge will be implemented in the recommendations to enable its application in the real conditions of the organisation.

Selected research findings:

- Employees rate the safety climate highly – the highest rated dimensions are the level of safe behaviour, the level of management commitment to H&S, technical facilities and ergonomics and the H&S training process. The lowest (but still within the area of high scores) are the pace of work and the level of fatigue and employee participation in H&S.

- The analysis showed that the organisation is not homogeneous in the assessment of the safety climate – the type of position held and the area of employment differentiate the perception of the safety climate.
- Employees declare pro-safety attitudes. Of the three attitude components, the highest scores are found in the cognitive aspect, meaning that employees are well informed about risks and hazards and have knowledge of how to protect themselves against them.
- Similar to the assessment of the safety climate, there are differences depending on the area of employment and the position held, which means that the organisation is not homogeneous with regard to the level of attitudes towards safety.
- Employees of the surveyed organisation are characterised by a high level of personal safety culture.
- There are statistically significant differences in the level of the individual's personal safety culture manifested in the professional sphere compared to the non-professional sphere, in favour of the professional sphere. The highest results occurred in the dimensions of ways to protect oneself from hazards, attitude to physical risks and care for environmental safety. This implies that the actions taken by organisations, including the procedures and standards implemented, effectively translate into employees' behaviour at work.
- An analysis of intergroup differences by area of employment showed that employees in the Administration area rated themselves lowest, while employees in the Health and Safety at Work area rated themselves highest.
- The analysis of intergroup differences taking into account the position occupied showed that persons holding managerial positions are characterised by a significantly higher overall level of the individual's personal safety culture and a significantly higher level of the individual's personal safety culture manifested in the professional and non-professional spheres compared to the other positions analysed, which is a very good result in the context of the scope of their tasks and duties – the manager models the behaviour of employees.

The research carried out, the analyses of the results obtained and the conclusions drawn from them made it possible to develop programmes:

1. *Programme for Building a Positive Safety Climate in the Organisation*, which included, among other things, proposals for improvements in the areas of strengthening the role of the health and safety services (integrating the activities of the health and safety services into the holistic management of the organisation), communication, reducing the distance, improving the participation procedure, diagnosing training needs in the context of creating pro-safety attitudes, integrating

topics related to the psychology of safe behaviour into training programmes and incorporating aspects related to monitoring attitudes towards safety into the recruitment process.

2. *A Programme for the Development of a Culture of Personal Safety* including issues related to building a community with a clear emphasis on the values that unite it, a proposal to conduct community-building activities in the area of assistance and cooperation with the external environment.

The findings of the research were presented in workshops with management and change leaders in the organisation. A workshop was also held to help develop the foundations for the most optimal implementation resulting from the research and to establish the timetable and scope for implementation. Leaders were sensitised to possible resistance to change and the need to take action to overcome it.

2 Behaviour-Based Safety
Theory and Practice

The contemporary approach to health and safety management, based on building a high health and safety culture, focuses on promoting informed, safe attitudes and behaviours among employees. The concept of Behaviour-Based Safety (BBS) provides a practical tool to support the development of a safety culture through the systematic observation, analysis and reinforcement of positive habits. BBS does not fit into standard safety programmes that have a beginning and an end; rather, BBS is an ongoing, never-ending process of building an H&S culture based on values, employee involvement, effective communication and positive reinforcement. The purpose of this chapter is to present the theoretical foundations of BBS, the ideas and stages of implementation and to highlight the key barriers and pitfalls in implementing and operating a BBS process. The chapter also presents the measures of the process and describes the practical aspects using the example of a selected organisation.

2.1 THE ESSENCE OF BEHAVIOUR-BASED SAFETY (BBS) – THEORETICAL ASSUMPTIONS AND STAGES OF IMPLEMENTATION

At the core of BBS is the assumption that workplace safety is based on a combination of three measurable factors: the employee, the environment and employee behaviour (McSween, 2003). In the case of the employee, at the recruitment stage, it is possible to select personnel based on physical conditions, work experience, qualifications and competencies. Similarly, we can influence and shape the environment, that is, the physical environment, technical environment, tasks assigned and health and safety culture. A critical element in achieving occupational safety and health (OSH) objectives, which is essentially beyond influence, is the employee and their behaviour. This issue is addressed by the BBS concept. A key populariser and theoretician of the BBS concept is the American psychologist E. Scott Geller (Geller, 2000). The BBS concept is derived from an interdisciplinary combination of behavioural psychology, organisational science and workplace safety management. It is based on the classic mechanism of instrumental conditioning, ABC (Antecedent-Behaviour-Consequence), developed by Skinner (Skinner, 1953). According to this theory, people's behaviour can be influenced by regularly rewarding desirable behaviour or by punishing rule-breaking. If an employee is rewarded for following a safety rule, they are more likely to repeat that behaviour, whereas if they are admonished over time, they should stop acting in a certain way.

DOI: 10.1201/9781003677741-2

The BBS process is based on the idea that workers' behaviour is not only influenced by their knowledge of hazards but also relies on changing their behaviour in daily duties. A key tool in this process is the systematic monitoring of behaviour and the provision of feedback. By observing behaviour, both positive and negative actions can be identified so that they can be further reinforced or eliminated (Krause, 1996). An important element in the successful implementation of a BBS is the involvement of all employees, including both management and line staff. Only shared responsibility and collaboration can bring about lasting change in the safety culture of an organisation and change employees' thinking that safety is a shared value (Carra et al., 2023; Cooper, 1998; Tuncel et al., 2006; Weaver et al., 2023).

2.1.1 Behaviour-Based Safety – As a Process

BBS should be seen as a continuous, never-ending process rather than a one-off H&S programme or intervention. It is not an ad hoc tool, but a complex and dynamic mechanism – its effectiveness is based on the systematic implementation, monitoring and improvement of activities involving workers at all levels of the organisation. Table 2.1 shows the main elements of BBS postulated by selected authors.

TABLE 2.1
Elements of the BBS Process

Author	Elements of the BBS Process
Terry E. McSween (McSween, 2003)	1. Management support 2. Defining safe behaviours 3. Establishing an observation system 4. Data analysis and feedback 5. Continuous improvement
E. Scott Geller (Geller, 2005)	*DO IT* model 1. Define (Dream/Operationalise) – define critical objectives of observed behaviour 2. Observe – observe, collect baseline data 3. Intervene – intervention and feedback 4. Test – repeat observation and evaluate effectiveness
Krause (Krause, 1996)	1. Planning and organisational preparation 2. Training and preparation of observers 3. Conducting the observation 4. Feedback 5. Data analysis and reporting
Al-Bayati A. J., Abreu, T. T. P. D. & Ko, C. H. (Al-Bayati et al., 2025)	1. Observation 2. Unsafe behaviour identification 3. Intervention 4. Evaluation

Source: Own elaboration.

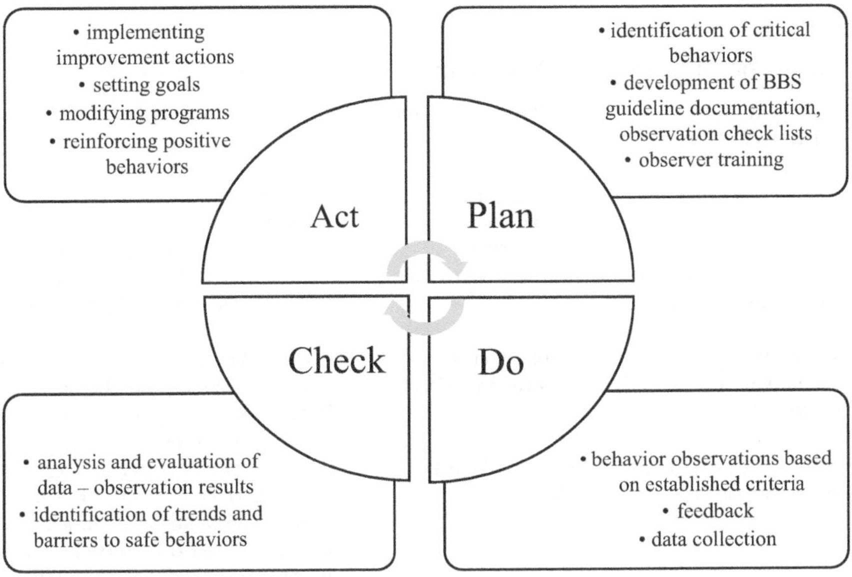

FIGURE 2.1 The BBS process as a PDCA Deming continuous improvement cycle.

Source: Own elaboration.

In the classifications of aforementioned BBS elements, all authors emphasise that it is a continuous process. It is, therefore, legitimate to compare the BBS process to Deming's PDCA continuous improvement cycle, as both approaches are based on systematic observation, data-driven adaptation of actions and continuous improvement in health and safety management (International Organisation for Standardization, 2018). A comparison of the BBS process to Deming's PDCA cycle is illustrated in Figure 2.1.

The key elements of the BBS process find their counterparts in an occupational safety management system based precisely on the PDCA cycle, that is, the requirements of ISO 45001 (International Organisation for Standardization, 2018), as shown in Table 2.2.

While ISO 45001 provides a formal framework for systemic health and safety management, the BBS behavioural observation process can be seen as an operational tool to support the implementation of ISO 45001, especially in the areas of behaviour, commitment and building a safety culture.

2.1.2 Behaviour-Based Safety – Phases of Implementation

As mentioned earlier, for BBS to have a chance to function properly, it is necessary to first ensure that the implementation process is carried out correctly and then to supervise its implementation effectively. The BBS process is a group process, and its size increases steadily with the expansion of areas and the rise in observation rates,

TABLE 2.2
Comparison of ISO 45001 Requirements and BBS Process Elements

Requirement of the Standard ISO 45001	BBS Process Element	Description
Point 5.1 – Leadership and commitment Point 5.4 – Consultation and participation of workers	Leadership and employee involvement	Both concepts require active employee participation at all levels of the organisation and support from top management. BBS emphasises positive reinforcement, which supports building the OSH culture required by ISO.
Point 6.1 – Hazard identification and risk assessment	Identification of critical, hazardous behaviour	The standard requires the identification and assessment of risks; the BBS identifies behaviours (including risky behaviours) as sources of hazards and risks.
Point 6.2 – H&S objectives and action planning	Improving the BBS, H&S objectives	The results of the observations (e.g. % of risky behaviour) can, at the same time, be indicators of the achievement of OSH objectives. The BBS observations are used to plan training and organisational changes, allowing changes in behaviour to be tracked; that is, the BBS supports the achievement of OSH objectives and allows them to be monitored in accordance with ISO.
Point 7.2 – Competencies	Training of observers	The observers of the BBS have to be trained, which is in line with the ISO requirements of ensuring adequate competence.
Point 7.4 – Communication	Feedback	A key element of the BBS is open, frank communication at the level of internal communication in the ISO standard.
Point 8.2 – Emergency preparedness and response	Intervention and non-routine intervention	In BBS, the observer intervenes immediately in the event of a life or health-threatening situation that is not included in the observation card and is not directly related to the position observed. ISO requires the development of procedures for unusual, emergency situations.
Point 9.1 – Monitoring, measurement, analysis and evaluation	Monitoring and analysis of observation data	ISO requires an analysis of the effectiveness of the system; in BBS, the process is analysed.
Point 10.2 – Non-conformities and corrective actions Point 10.3 – Continuous improvement	Corrective action and improvement	Both concepts imply corrective action and continuous process improvement.

Source: Own elaboration.

reaching up to 100% of the workforce. The respective groups, in their respective phases, carry out specific tasks, as summarised in Table 2.3.

The names given above, as well as the sheer number of teams in practice, may vary. For the BBS process itself, this is of little relevance, as long as specific action steps are followed, the appropriate procedures are developed for them and the teams are trained properly.

TABLE 2.3
Implementation Phases of the BBS Process and Team Responsibilities

Teams and Their Composition	Responsibility of the Teams
Design phase	
Management team – top management	- making key decisions for the process (e.g. areas covered by the BBS process, type of overt/hidden observation, the number of qualified observers); - setting objectives together with measures for achieving these objectives; - approval of process implementation plans; - communication of the project within the organisation; - constant cooperation with the implementation team.
Implementation team – workers of health and safety departments, health and safety specialists, representatives of engineering and technical staff, area managers, leaders, selected staff representatives and external experts (in case the implementation is carried out by an external company) – about 7–10 people	- preparation of a boundary framework for implementation, analysis of risks and opportunities for the project; - creation of implementation plans for the BBS process; - development and preparation of procedures (e.g. operation of the process, training of qualified observers, conducting observations and providing feedback, recruitment of qualified observers, reporting and verification of observations, taking corrective actions); - development of an implementation schedule with resources and tasks and responsibilities; - development of observation cards based on risk analysis and identified critical/risk behaviours; - recruitment and training of observers; - development of information and training materials; - development of reporting rules.
Implementation phase	
Management team – top management	- monitoring the progress of the BBS process, - auditing the BBS process, - taking corrective actions, - setting directions for improvement of the BBS process, - cooperating with the coordination team on a continuous basis, - initiating discussions and ensuring understanding of the safety issues within the company, - motivating qualified observers and members of the organisation to actively participate in the process.

(Continued)

TABLE 2.3 (CONTINUED)
Implementation Phases of the BBS Process and Team Responsibilities

Teams and Their Composition	Responsibility of the Teams
Coordination team for the implementation of the BBS process – H&S staff, managers, representatives of the maintenance department, representatives of the human resources department, representatives of the top management	- making decisions on the course and development of the process, analysing its implementation and making adjustments related to the course of the process, - collecting information from observations, - setting corrective objectives, - taking day-to-day corrective actions, - taking ongoing process improvement actions, - communicating with qualified observers in order to analyse the processes and difficulties that may arise during the observation, - providing support to qualified observers adequate to the reported problems and needs (e.g. organising additional training, coaching, mentoring of qualified observers), - recruiting new observers, - analysing observation data indicators of the observation process, - monitoring the number of observations carried out, - drafting conclusions and reports on the process, - communicating observation results to top management and staff, - developing proposals for incentive measures for qualified observers, - developing corrective actions for management and improvement plans at strategic level.
Executive team – qualified observers	- carrying out observations, - providing feedback, - intervening in situations of imminent danger to life and health, - completing the observation card, - reporting the results of the observation.

Source: Own elaboration based on McSween (2003).

2.1.3 BBS Implementation Schedule

Adequate time needs to be allocated for the implementation process to be carried out correctly and without undue stress. The length of the implementation process depends on many factors, including the size of the plant, the area in which the process is to be implemented, the specific organisation of the work in terms of the variety and complexity of the activities to be performed and the time capabilities of the implementation team. Optimally, an organisation should have between 6 and 12 months at its disposal. An extremely important stage is the proper planning of the BBS implementation process in terms of time and resource realities. A schedule not only helps to clarify the scope of activities and the dependencies between them, but also facilitates supervision and early detection of risks in the implementation of the various stages of the project. Nowadays, project management computer applications are proving particularly useful. One of the most popular is Microsoft Project, which, thanks to its functionality in the form of numerous reporting functions and extensive interfaces for the management of important aspects of the project, such as the management of resources, time and finances, allows the implementation process to be constantly monitored and the schedule to be optimised. The preparation of the schedule in a graphical version can be carried out using tools such as IC Project, ProjectLibre, ClickUp, Canva, Capterra or standard MS Office software.

2.1.4 Basic Principles of the BBS Process

As a first step, top management, in cooperation with the implementation team, should make key decisions on the assumptions of the BBS process by answering some basic questions:

- In which area will the observation be carried out (e.g. production, warehouse, whole plant, one department)?
- Which employees will be observed (e.g. all working people, only full-time employees)?
- What type of observation will it be (overt or covert)?
- What will the frequency of observation in the area be (e.g. more frequent in accident-prone areas)?
- What will the number of qualified observers be?
- What will the number of observations carried out by a qualified observer be? The next step is to define the rules for recruiting, training, motivating and supporting observers and to determine where and in what form observers collect observation cards and where they submit them, as well as the rules for reporting observation results.

The adoption of the initial assumptions of the BBS process is reflected in the document *Procedure/Instruction of the BBS Process*, which, along with its annexes, contains the basic guidelines and rules for implementing the process, as adopted and approved during its implementation. It can be modified at any stage of the process. The *Procedure/Instruction of the BBS Process* should include standard system elements such as scope, basic definitions, process flow and description, responsibilities and tasks to be performed in each sub-process.

The planned BBS process depends on the specifics of the organisation, its resource capabilities and the H&S objectives set for implementation. It is important to remember that the more the process is "tailored" to the needs of the organisation, the greater its effectiveness.

2.1.5 Pre-audit – Identification of Risk Behaviour

Before starting the implementation work, the implementation team needs to initially identify emerging risk behaviours in the organisation. A valuable source of information is the post-accident documentation (identified causes of occupational accidents), near misses (if such records are kept in the organisation), the results of inspections by state labour inspection bodies, the results of internal inspections by OSH staff, the results of internal and external audits or observations carried out in the company and interviews with employees.

Interviews with middle and senior management and OSH workers do not always provide an accurate picture of the situation, as they often engage in a kind of self-assessment of their actions, fail to objectively evaluate the circumstances, downplay the scale of negligence and irregularities and utilise the rhetoric of success. A key source of information, however, is the interviews conducted with rank-and-file employees. It is recommended that such interviews be carried out by individuals who are not the direct superiors of the employees. Openness and a lack of fear among employees can be assured by an external expert. Interviews with employees should be conducted in small groups (a maximum of several hundred people), without the presence of management. The process of collecting information from rank-and-file workers also presents a good opportunity to explain (to workers) the essence and objectives of the planned implementation of the BBS process, and the act of listening to the voice of workers is a form of participatory OSH management (Boczkowska et al., 2022).

This kind of audit, that is, the combination of knowledge from all sources of information (from documents, through interviews and observations) about hazards occurring in the organisation, dangerous situations and risky behaviour, allows the development of a list of critical/risky behaviours in specific areas (e.g. the external area of the company, the production area, the warehouse area), which will form the basis for the development of observation cards and the training of future observers.

2.1.6 Observation Cards

The basic tool of a qualified observer is the *observation card*, which is difficult and time-consuming to develop. Some implementation teams choose to develop a single unified observation card (Fazilov et al., 2018), while others prepare several observation cards after analysing the workplaces in advance and grouping them according to the specifics of the work and existing hazards at the workplaces (Nunu et al., 2019). Undoubtedly, the advantage of a single observation card is the short time needed to develop it and its universal nature, requiring no expertise from the qualified observer, which simplifies the work, and one large database is available to us in the process of BBS measurement and monitoring. However, the generic nature of the observation card will also result in conclusions drawn from observations of employee behaviour. Thus, inappropriate or risky behaviour specific to a particular job will not be identified and eliminated. Observation cards dedicated to a specific job group require more effort at the stage of preparation and training of observers. However, they allow for more in-depth observations and significantly increase awareness of hazards and risks both within the group of observers and, after the intervention, within the group of workers observed, thus enabling the change of behaviour from risky to safe.

The structure of the observation card can vary depending on what information is needed to monitor the BBS process and build appropriate metrics. It should include details of the observer, the time and place (position, department) of the observation. The structure of the Charter should be well thought out, easy to read and easy for the observer to fill in. It is advisable to create observation cards no longer than one sheet (maximum two pages). Creating longer observation cards will make the observation process itself much more difficult and costly.

The main part of the card is a list of critical employee behaviours to be assessed by a qualified observer. Once the list of critical activities has been created, the behaviours to be observed can be subdivided into subgroups of so-called "categories" concerning, for example:

- personal protective equipment used,
- tools and equipment used,
- compliance with company rules and procedures,
- compliance with work discipline, working posture and ergonomics.

Critical activities should then be formulated in terms of safe rather than risky behaviour – for example, "The worker uses ear protection in a designated area". This type of wording is crucial because firstly the qualified observer needs to know what type of behaviour is safe; secondly, the idea of the BBS is to focus on and reinforce the positive aspects of behaviour. Negative wording will result in the observer's attention being diverted from the positive to the negative aspects of the behaviour, which – in practice – may translate into more attention being given to the identified risky behaviours to the exclusion of positive feedback for those behaviours that were safe.

In the observation card, the qualified observer identifies all behaviours, both safe and risky. Both types of behaviour are equally important to the success of the process: safe – to reinforce these behaviours; risky – to find out from the person displaying these behaviours why they are taking these risks.

As observations are carried out randomly, at different times and stages of the work process, the qualified observer will not always witness all the behaviours listed on the observation card. Therefore, it is important that, in addition to a positive ("+") or negative ("–") rating, it is possible to select the option *not applicable* to ensure that all fields on the sheet are completed fully. The qualified observer is obliged to intervene immediately in a situation of imminent danger to life and health or to undertake so-called "non-standard intervention" after identifying other dangerous behaviours or situations not included in the observation card and not directly related to the observed site. These situations should be documented in the observation card; therefore, when preparing the card, it is advisable to leave space for *other comments and observations from the observer.*

Once the observation card has been approved by top management, sample observations are made (by one or two observers at a time) and any adjustments to the card are implemented.

2.1.7 Qualified Observer

In the BBS process, the qualified observer plays a key role. They have direct contact with the workers. They not only collect the data gathered during the observation but also interview the observer. The quality of the observation process depends on their preparation, knowledge, experience and personal and social competencies.

A qualified observer can be any person who occupies a parallel position to the person being observed; it is optimal for qualified observers to be rank-and-file employees.

The selection of staff observers for the BBS process should be preceded by a verification of the competence of the candidates. Since the key element is properly conducted feedback, the qualified observer must possess key communication competencies. They should be perceptive, conscientious, systematic, objective, free of bias and, above all, convincing. They should be knowledgeable and experienced, credible, open-minded, objective and non-confrontational. Most important, however, is the individual's personal motivation to participate – an employee forced into the role of observer by a superior will not be a good observer.

Those selected for the role of observer must receive appropriate training. Its purpose is to learn observation skills with positive feedback in the BBS process. The training is based on a pre-developed programme, which covers aspects such as safety culture management, goals and objectives of observation, types of behaviour to be observed, how to communicate and reporting, among others. As a result of the training, candidates for observers should have knowledge in the following areas: the mechanisms and motivation to engage in risky behaviour, the conditions that must be met to enable safe behaviour, the specifics of the work and risks and the principles of the BBS process in an organisation (in particular, the principles of observation, feedback and reporting; Weaver et al., 2023).

The issue of observer motivation should be well considered by the organisation. Qualified observers can be motivated in a variety of ways: from giving positive feedback to observers on their performance in the BBS process, including this activity in the employee appraisal system (which does not generate additional costs), to various benefits or points that translate into rewards, such as the opportunity to go to the cinema, theatre or out to dinner.

However, it is worth emphasising that the process of recruiting observers continues throughout the implementation of BBS. This is due to two reasons: (1) employees rotate or do not perform the tasks imposed on them adequately, decide to resign from the process for personal reasons and an appropriate number of them must be secured; (2) often, at the implementation stage, an initial level of observers is set with the intention of increasing it. Therefore, recruitment of observers is also carried out during the implementation phase of the BBS process.

2.1.8 Observation and Feedback

A key element of the BBS concept is the observation of behaviour and the provision of feedback. Observations must target specific, measurable and observable behaviours that directly affect safety in the workplace. Weaver et al. (2023) indicate that the effectiveness of observations depends on the concrete definition of behaviours that are considered safe and unsafe. Only then will observers be able to assess which of the observed behaviours are likely to lead to accidental incidents so that similar situations can be prevented in the future.

The observation process itself will not work if employees do not receive properly formulated feedback. Feedback is a key element of the BBS process (Geller, 2000). The feedback given to employees influences their motivation to change their behaviour.

Giving feedback from an observation in the context of implementing a BBS process is different from a traditional employee appraisal. It is not only a matter of pointing out mistakes; positive behaviour should also be highlighted. Indeed, the information provided is a direct mechanism for reinforcing or correcting behaviour, encouraging employees to understand the consequences of actions performed and to make more informed decisions in the future (Komaki et al., 1980; Koźlik, 2008).

In the context of a BBS strategy, feedback should have a primarily motivational and educational function. Positive information contributes to the reinforcement of safe habits and enhances the employee's sense of safety influence (Daniels & Bailey, 2014). The information provided should be specific and precise. It should refer only to identified and directly observed behaviour, excluding generalisations, emotional judgements, personality judgements or insinuations about the employee's intentions (Daniels & Bailey, 2014; Znajmiecka & Boczkowska, 2017). Feedback provided in this way minimises the risk of defensive attitudes, avoids unnecessary conflict and helps the employee understand which behaviours to repeat and which need to be changed. The feedback given is intended to reinforce positive behaviour, so the information provided should not focus on criticism, but should also be positive, being supportive rather than corrective (Cooper, 2009). The principles of interviewing an observed employee are shown in Figure 2.2.

Various types of difficult situations also arise in the observation process. These may involve an inappropriate attitude of the immediate supervisor towards the observation process related to prioritising other tasks (production, quality) over it. Difficulties may also arise between the qualified observer and the observed person, who may react negatively to the observation process. The qualified observer may also need support in identifying risks or in interviewing the observed person. In the context of the multitude of problems that may arise in the observation process, ongoing contact with qualified observers is of particular importance. It is essential that meetings with them are conducted systematically and include an analysis of the problems that arise and allow them to practise the competencies for solving these issues. Tools such as mentoring or coaching can also be used to support qualified observers.

2.2 IMPLEMENTING, MAINTAINING AND IMPROVING BBS – BARRIERS AND PITFALLS

The implementation of BBS processes faces a number of barriers that can significantly determine the success of the implementation. The key barriers are primarily as follows (Choudhry, 2014; Galloway, 2011; Lundell & Marcham, 2018; Rajput & Makwana, 2025; Thakare & Singh, 2022):

- *Lack of top management commitment and support* – the implementation of a BBS requires the active involvement of top management and cannot be treated as solely the opinion of the H&S department or the H&S coordinator. Without the visible involvement of top management, the BBS process quickly becomes irrelevant. Managers should personally model safe behaviour (use of PPE, adherence to procedures); if this is not the case, workers see a gap between declarations and practice, that is, a straightforward disavowal

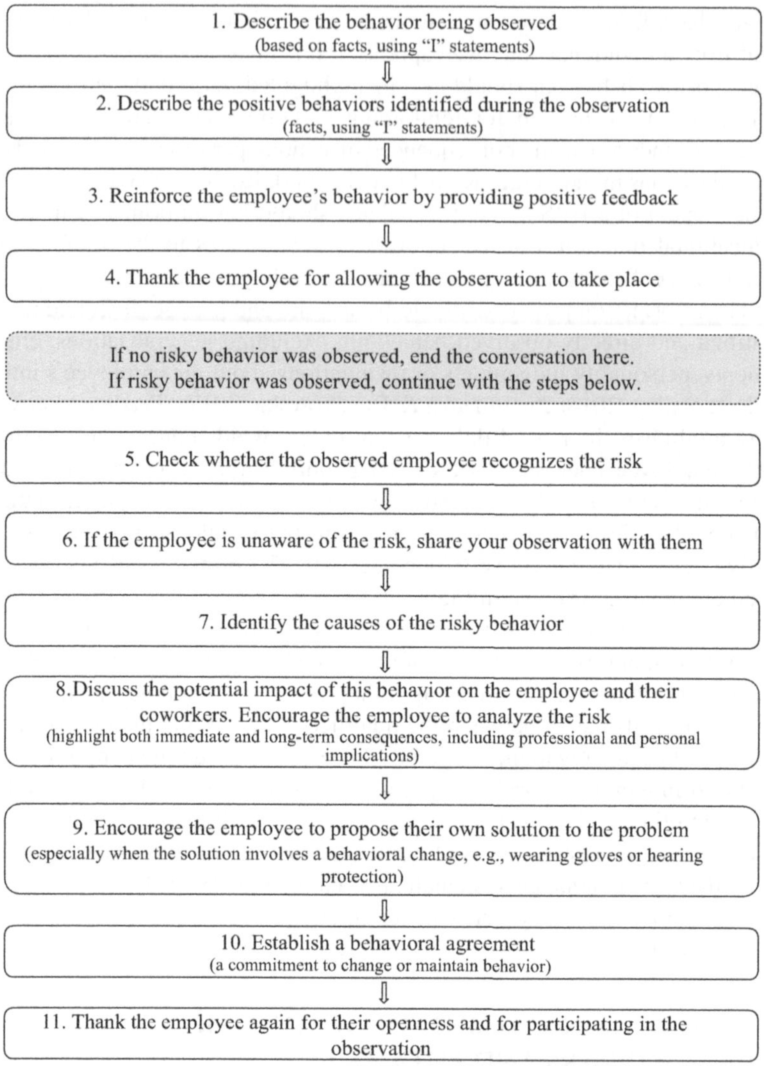

FIGURE 2.2 Principles for qualified observer feedback in the BBS process.

Source: Own elaboration.

of the BBS process by management. If, despite the systematic collection of observational data, no corrective action is taken by management, rank-and-file workers see the process as ineffective and lose motivation to continue.

- *Low level of acceptance of the BBS process by rank-and-file employees* – standard employee resistance to change due to lack of trust or fear of new procedures requires an appropriate communication strategy. Implementing a BBS without a clear rationale and explanation of the rules can lead to low employee engagement and, therefore, failure of the process.

- *The risk of a "blame culture"* – the BBS process and the implementation of observations must not be seen as a "blame search" for inappropriate, undesirable behaviour. The essence is to look for and analyse systemic causes and improve processes on the basis of positive feedback. Disciplinary measures cannot, therefore, be applied to employees, as this may end up hiding mistakes and eroding trust. BBS and identified risky behaviour must be treated as an opportunity to learn and improve the process. An atmosphere of fear and mistrust stifles the active participation of workers in improving workplace safety.
- *Lack of resources* – the BBS process requires resources: human – BBS leaders and coordinators, trained observers to carry out observations and to provide constructive feedback and involvement from top management; organisational – teams, procedures and documents; technical – software and IT infrastructure; time – time for observations, data analysis, action planning, training and coaching. Failure to adequately plan and ensure continuity in the implementation of the BBS process leads to a fading of the initiative.

It is important to realise that a completed implementation process does not automatically mean success. There are also obstacles to overcome during the BBS operation and improvement phase. Barriers to maintaining and improving the BBS process are, in particular:

- After the implementation phase, there is often a decline in commitment – the initial enthusiasm of the participants may wane over time, especially if there is a lack of visible results and insufficient support from leaders, coordinators or inadequate change management in the organisation.
- Data collected during observations is redundant and not analysed by top management. Collecting data without analysing it regularly and translating it into corrective action results in a loss of confidence in the programme ("we collect but do nothing about it").
- Irregularity and loss of systematicity. When observations are not carried out systematically (e.g. due to staff shortages, tight schedules, prioritisation of production goals), the programme loses continuity. This can be exacerbated by a high turnover of staff or observers, as staff changes without adequate information and training of successors weaken the structure of the programme.
- Observer fatigue and burnout. Carrying out observations without recognition, rewards or opportunities for rotation leads to fatigue and abandonment. In addition, the psychological strain on observers associated with colleagues being critical of BBS and observing without adequate internal communication can accelerate the process.
- Poor feedback. Lack of conversations discussing performance with teams and lack of positive feedback demotivates staff and observers.
- Lack of integration with the safety management system – if BBS operates as a separate initiative rather than as part of an integrated health and safety system (e.g. ISO 45001), it is easily marginalised.

- Lack of development of the BBS process – a process that does not evolve becomes rigid and loses relevance and importance. The organisation needs to learn lessons, improve processes, update documentation (e.g. observation cards, BBS procedures) and set new goals.

Despite the apparent success, mistakes or inefficiencies can occur during the implementation, then operation and improvement of the BBS process that lead to the distortion of the BBS concept or inhibit the effectiveness of the BBS process. The most important pitfalls in the implementation and operation of a BBS scheme can include:

1. *Apparent implementation savings* – the implementation of a BBS scheme may be carried out "at no cost" to the organisation's staff. However, the savings may turn out to be apparent due to the need for additional training, lack of staff experience in BBS implementation and a greater risk of mistakes being made, thereby increasing implementation time. Benchmarking and the objectivity of external experts provide a greater guarantee of a smooth BBS launch.
2. *Expecting immediate results* – failure to improve accident rates may lead to abandoning BBS too early. It is important to remember that quantitative indicators must be preceded by a change in H&S behaviour and culture, which takes time. In the first instance, positive effects of implementation should be sought by monitoring qualitative effects (hazard reports, safety conversations and H&S culture). Pressure for quick results is not conducive to achieving sustainable results and can lead to lower engagement.
3. *Inadequate feedback skills* – observers without solid training in communicating difficult information cause fear and resistance among employees. Furthermore, the subjectivity of evaluations and discrepancies in the interpretation of behaviour can reduce the consistency and reliability of data, which is why it is important to continuously monitor the quality of feedback.
4. *Assigned observers* – the lack of proper involvement of employees or the achievement of "unrealistic" goals in relation to the planned increasing number of observers can lead to supervisors typecasting specific individuals as observers. Employees who have been appointed against their will to carry out observations rarely duly fulfil their tasks, and the whole BBS process becomes a mystification – resulting in false data for the organisation. From the point of view of the effectiveness of the BBS process, it is more effective to recruit "volunteers" who are intrinsically motivated to take action to improve workplace safety in the organisation or for whom the incentive system developed for participation is sufficient to make the extra effort.
5. *Inappropriate reward and punishment systems* – the use of a system that rewards the number of observations completed without verifying their quality can result in poor data quality and manipulation of results. In addition, the use of excessive economic motivators (financial rewards) can undermine the intrinsic motivation of employees and lead to sham engagement behaviour.

6. *Extensive documentation* – overly elaborate forms (observation cards) can be problematic for observers and can lead to the collection of irrelevant data, generating unnecessary wasted time.

2.3 MEASURING AND MONITORING BBS

Kaplan noted that measurement is crucial in management: "if you can't measure something, you can't manage it" (Kaplan, 1996). Measurers are like "sensors of an organisation or process"; they should be linked to the company's objectives at all levels and, as far as possible, reflect the relationships between them.

In the context of occupational health and safety, measurement can be based (Bayramowa et al., 2023) on:

- lagging indicators (also known as outcome indicators or negative performance indicators) – which generally focus on measuring losses resulting from poor working conditions, for example, frequency of occupational accidents, number of days of absenteeism due to occupational accidents, incidence rate of occupational diseases, costs of occupational accident compensation paid.
- leading indicators (activity indicators or positive performance indicators) – which characterise the prevention potential of the companies, as well as the activities carried out to improve the safety and health protection of employees, measures that predict or signal the risk of incidents or accidents before they occur, for example, the rate of at-risk employment, the safety climate index, the level of OSH worker participation, the number of OSH audits carried out, the number of near misses, the number of observations of safe/hazardous behaviour.

In the context of monitoring the BBS process, the construction of indicators requires knowledge based on reliable, quantitative information. The key, of course, is the benchmark – that is, the defined levels of achievement of specific tasks. Properly defining and setting metrics provides the basis for further analysis, assessment of BBS performance and improvement.

The following are examples of leading indicators used to monitor and improve the BBS process in the short or long term.

1. *Schedule Adherence Indicator* – indicates the percentage (%) of activities or milestones that have been completed according to the BBS implementation schedule:

$$ATI = \frac{NAC}{NAP} \times 100\% \qquad (2.1)$$

Where:

ATI – activity timeliness indicator
NAC – number of activities completed on time
NAP – number of activities planned to date

Typically, a level of ATI at 90% is considered satisfactory, indicating very good compliance with the schedule; 70–89% represents moderate compliance requiring an analysis of the reasons for delays, and, if ATI < 70%, corrective action must be taken as the effective implementation of BBS is at risk.

2. *Observation reliability index* – the creation of an ideal observation card is a process that is carried out on an experimental basis, often even by trial and error. Trial observations conducted by two observers simultaneously can provide valuable information. Once the observation data from both cards have been collected, the so-called "Observation Reliability Index" can be determined, which indicates the percentage of concordant responses in relation to all observations. The Observation Reliability Index should be 80% or above (McSween, 2003):

$$ORI = \frac{NMO}{NMO + NDO} \times 100\% \tag{2.2}$$

   Where:
   - ORI – observation reliability index
   - NMO – number of matching observations
   - NDO – number of discordant observations

3. *Observability rate in an organisation/area* – an indicator showing the percentage of posts covered by the BBS process. Typically, the main, often long-term objective of BBS is to cover the whole organisation or a selected area of it, that is, a specific closed part. Achieving the set target of 100% of posts being assessed requires the involvement of a large group of qualified observers, which is often not feasible at the start of BBS. The authors' experience suggests that a minimum of 20% of observable posts should be adopted at the outset and then increased as the BBS process continues. In some organisations, up to 100% has already been achieved, meaning that every employee is subject to observation within a month.

   To monitor this process and clearly indicate the targets, the so-called "observability index" can be used:

$$OI = \frac{NOE}{NE} \times 100\% \tag{2.3}$$

   Where:
   - OI – observability index
   - NOE – number of employees/positions covered by the observation
   - NE – number of all employees/positions in the organisation or area covered by the BBS process

It should be noted that the value of the observability rate OI can vary for different groups of jobs. If one wishes to quickly eliminate the most accident-prone areas, for example, in the warehouse where the accident statistics are much higher than in production, one can assume a higher observation intensity (more positions covered by BBS) in the warehouse than in other areas.

4. *Average of identified behaviours* – an indicator showing the average number of all behaviours identified during one observation. During the observation process, the qualified observer has the option to leave certain behaviours unassessed by writing *not applicable* on the observation card. As the BBS process develops, qualified observers gain experience, their observations become more insightful and they demonstrate greater perceptiveness. An increasing value of the above indicator can measure the growing competence of the observer and the quality of the observations conducted. A decreasing value of the indicator may signify a deterioration in the quality of a qualified observer's work and should prompt corrective action in the observation process, which involves additional training for observers. This measure is used in the individual assessment of observers' work:

$$\mathrm{AOB} = \frac{\mathrm{NOB}}{\mathrm{NO}} \tag{2.4}$$

Where:

AOB – average of behaviours identified during one observation
NOB – number of all behaviours identified during all observations in a given period
NO – number of completed observations in a given period

It should be borne in mind that during an observation, not all the behaviours subject to assessment by the observer need to occur; those that do not may be marked as *not applicable*. Therefore, caution should be exercised in the analysis of this indicator.

5. *Qualified observer activity indicator* – allows monitoring the level of activity, involvement of qualified observers globally, that is, for the whole BBS process, as well as individually for each observer:

$$\mathrm{AQOI} = \frac{\mathrm{NO}}{\mathrm{NPO}} \times 100\% \tag{2.5}$$

Where:

AQOI – activity index of qualified observers
NO – number of observations made during the period
NPO – number of planned observations in a given period

The above indicator can take values both lower and higher than 100%. Obviously, the recommended target is to achieve a level of 100% of planned observations in all areas covered by the observation process and for all qualified observers. However, practice shows that a satisfactory – minimum – level of the observation rate, taking into account observer absenteeism, random situations, holidays, and so on adopted in most organisations, is 75%. The authors' experience shows that planning training for a larger number of qualified observers during the implementation stage is worthwhile, as it is rare for all observers to achieve the planned 100% observation rate. Therefore, this rate should be correlated with the observability rate OI (Equation 2.3). In order to maintain the assumed level of the index OI, it is, therefore, recommended to train 10% more observers than the basic calculation would suggest.

6. *Reliability index of observation cards* – the implementation of systematic observations should be subject to a reliability assessment. In practice, it often occurs that the cards filled in by qualified observers are incomplete and contain numerous gaps, which makes the observation results unsuitable for further analysis or may generate erroneous conclusions. Therefore, in monitoring the BBS process, a yardstick can be used to assess the performance of qualified observers:

$$\mathrm{ROCI} = \frac{\mathrm{NMFOC}}{\mathrm{NFOC}} \times 100\% \tag{2.6}$$

Where:
ROCI – reliability index of observation cards
NMFOC – number of missing, unfilled fields in observation cards
NFOC – number of all fields in observation cards required to be filled in

Of course, depending on the design of the card, the fields required to be filled in may include not only the assessment of the behaviour of the employees under observation, but also fields identifying the area under observation, the date of observation, the shift (in shift mode), the job title, and so on.

7. *Indicator of qualified observers* – as mentioned many times, the implementation of the BBS process takes time, in particular the preparation of an adequate number of qualified observers. This number results from the assumptions made, that is, the size of the area covered by the process, the observability rate OI (Equation 2.3) and the assumed level of observations performed by one qualified observer – the rate AQOI (Equation 2.5). Usually, in the initial phase of BBS implementation, the group of qualified observers is small and, as the process develops and improves, the number of qualified observers increases, reaching, in exceptional cases, even 100% – which means that all employees become qualified observers. The qualified observer indicator can be used to monitor this process:

$$\mathrm{QOI} = \frac{\mathrm{NQO}}{\mathrm{NE}} \times 100\% \tag{2.7}$$

Where:
QOI – qualified observer index
NQO – number of qualified observers
NE– number of all employees working in the organisation or area covered by the BBS process

Achieving a level of QOI = 100% brings with it a very important, even crucial aspect. The very fact of being a qualified observer naturally forces a change in the observer's attitude in their daily work. Qualified observers, through active participation in BBS, gain more knowledge and experience; thus, adherence to H&S regulations and rules and renunciation of unsafe behaviour become standard for the qualified observer.

8. *Safe Behaviour Index* – a key aspect of creating and reinforcing safe behaviour is to monitor behaviour and analyse its level, and this is served by the Safe Behaviour Index:

$$\mathrm{SBI} = \frac{\mathrm{NSB}}{\mathrm{NSB} + \mathrm{NRB}} \times 100\% \quad (2.8)$$

Where:
SBI – safe behaviour index
NSB – number of safe behaviours during observation
NRB – number of risky behaviours observed

The analysis of the level of the safe behaviour indicator can be carried out globally for the entire organisation included in the BBS process or for its separate areas or departments. The use of this indicator for comparison purposes or to study the dynamics of change from one period of BBS operation to the next is regarded as superior to the previously presented metrics.

9. *Indicator of risky behaviours* – analogous to the previous indicator, it is possible to estimate the level of negative behaviours identified during observations, expressed by the indicator of risky behaviours:

$$\mathrm{RBI} = \frac{\mathrm{NRB}}{\mathrm{NSB} + \mathrm{NRB}} \times 100\% \quad (2.9)$$

Where:
RBI – index of risky (hazardous) behaviour
NSB – number of safe behaviours during observation
NRB – number of risky behaviours during observations

The relationship between the rate of safe and risky behaviours is shown by the following equation:

$$\mathrm{SBI} + \mathrm{RBI} = 100\% \quad (2.10)$$

10. *Indicator of non-standard interventions* – the task of the qualified observer during the execution of the observation is also to identify other unsafe behaviours and situations that are not included in the observation card and are not directly related to the observed position (e.g. an unsecured forklift truck observed standing in the way of the position, transport paths noted to be obstructed). It is the duty of the observer to make a non-standard intervention and to note this fact in the observation card in the field – *other comments and observations from the observer* – and to inform the supervisor of

the given area about this fact. Such situations should also be monitored and evaluated, and the following indicator of non-standard interventions relating to identified so-called "other risks" may be useful:

$$NSII = \frac{NNSI}{NO} \times 100\% \tag{2.11}$$

Where:

NSII – non-standard intervention indicator
NNSI – number of non-standard interventions undertaken
NO – number of observations carried out

Depending on the level of achievement of the objectives defined by the above process metrics, corrective action needs to be taken or new ambitious targets set as part of continuous improvement. This can be carried out according to the guidelines presented in Table 2.4.

The presented set of indicators is not a closed catalogue; it can serve as a starting point for other, more in-depth analyses. Depending on the advancement of the BBS process and the availability of data, other metrics can be constructed. However, it is

TABLE 2.4
Actions Taken in the Context of the Values of the BBS Process Measures

Metre	Action to be Taken	
	Corrective Action	Improvement Actions-New Targets
Observability indicator in the organisation/area – OI	When OI does not achieve the target values, e.g., OI <20%	Increase in OI (max. 100%)
Average of identified behaviours – AOB	—*	Increase in AOB
Qualified observer activity indicator – AQOI	When AQOI does not reach the target values, e.g., AQOI <75%	Increase in AQOI (max. 100%)
Reliability index of observation cards – ROCI	When ROCI does not reach the target values, e.g., ROCI<90%	Increase in ROCI (max. 100%)
Qualified observer index – QOI	—*	Increase in QOI (max. 100%)
Safe behaviour indicator – SBI	—*	Increase in SBI (max. 100%)
Indicator of risky behaviours – RBI	—*	Decrease in RBI (min. 0%)
Non-standard intervention indicator – NSII	—*	Decrease in NSII (min. 0%)

* — at the launch/implementation stage of the BBS process, estimation of the indicator value is not possible; in subsequent improvement cycles, it may take the values achieved in the first cycle.Source: own elaboration

worth remembering that the main idea of the BBS process is observation and positive reinforcement, and that an excessive number of indicators or elaborate statistical analyses may obscure this idea. It is also worth emphasising that the analyses and evaluations of the BBS process developed using the presented metrics should be correlated with the organisation's main H&S objectives. Sticking to the principle that each objective should be measurable, that is, quantified, it is worth evaluating the BBS process indicators alongside safety level metrics, including lagging indicators, which provide a snapshot of the scale of success or failure, that is, the effectiveness of BBS implementation. The effectiveness of BBS is evidenced by research reports, which in particular observed:

- a decrease in the rate of occupational accidents (Alavosius & Burleigh, 2021; Ayuni et al., 2022; Chen & Tian, 2012; Esitikot et al., 2025),
- a decrease in risky behaviour (Choudhry, 2014; Rashid et al., 2022; Yang, 2025; Yang & Kwon, 2022; Zakaria et al., 2023),
- an improvement in safety culture (Boczkowska & Znajmiecka-Sikora, 2015; Fatolitis, 2018; Zhang et al., 2023).

2.4 BBS EFFECTIVENESS IN PRACTICE – A CASE STUDY

2.4.1 BBS Implementation – Success Or Illusion?

The company is a recognisable brand in the domestic and international markets and has been in operation for more than 100 years. In the 1990s, it became part of a multinational corporation. It now employs over 500 people, of whom more than 400 are production workers, mostly men, with an average age of over 50 and seniority of more than 30 years. The workforce is highly integrated and finds it difficult to accept outsiders from the local community. The company's accident rate has been high for years, consistently above the national average, with an accident frequency rate of eight.

In 2019, the company hired a new H&S specialist, as the previous one had retired. The new H&S specialist, who was from outside the integrated community, was a young, ambitious individual eager to make a difference, demonstrate his proficiency and effect significant change in the organisation regarding H&S. He completed a three-day training course on BBS, which covered the process, implementation steps and operating principles. Fascinated by the concept and full of confidence, he was convinced that he could manage the implementation process independently, which would also be a cost-effective option for the organisation. The director accepted the proposal to implement BBS and gave the "green light" to the head of the H&S unit, who began to take action. As a first step, he drafted a general BBS Process Manual, modelled on training documentation, which addressed basic issues such as the openness of observations, frequency of observations per observer and the minimum number of observers. After consulting with the trade unions, the document was submitted to the director for approval.

In the next stage, the preparation of the observation cards was proceeded with. The health and safety officer organised meetings with the section managers and executive staff. Managers initially showed resistance to the change, as they thought that BBS being implemented would involve their additional involvement and additional responsibilities. The managers' calmness and acceptance came when they learned that the idea of BBS was to observe and provide feedback at the employee level (←→employee). Therefore, the managers' need to act on the BBS process and the proposed formula did not involve any burden.

Workers also reacted with anxiety at the beginning, fearing that they would be watched and that they would be looked for dirt on, which could result in penalties (e.g. no bonus). They did not want change, did not see the point and need for change in the area of H&S and feared additional responsibilities. Concerns disappeared when it became clear that the observations would be conducted openly, carried out by workers in parallel positions, without any negative consequences.

During the meetings, the employees were active, open, willing to identify hazards and eager to talk about what could be improved.

The head of H&S prepared a total of seven observation cards, which were accepted by the heads of the various departments and executive staff.

The next step was to recruit and train the observers. As planned, 20% of the crew came forward, motivated and encouraged by non-wage benefits. The observers were trained by the health and safety unit. The biggest problem proved to be giving feedback, especially positive feedback. Praising a colleague for working in accordance with the regulations, for using the assigned protective equipment and for maintaining order and tidiness on the job was the greatest difficulty, as workers do not know how to discuss the positives of the job; safety is only addressed when something is wrong. This novelty of "positive feedback" proved to be the greatest challenge.

In accordance with Point 6.2 of ISO 45001 (International Organisation for Standardization, 2018), the company has set targets for the next six months related to the BBS process (Table 2.5).

The observers proceeded to carry out the planned average of four observations per month with enthusiasm. After a "honeymoon month" during which everything worked as planned, the observers stopped delivering the observations, explaining that they did not have time to leave their workstation. Managers did not allow

TABLE 2.5
Objectives with Metrics for Achieving the BBS Objectives

Targets	Target Achievement Measure	
	Current Status	**Expected Value**
Main objective		
Reduction of the rate of accidents at work by 20%	$AI_{frequency} = 8$	$AI_{frequency} = 6.4$
Specific objectives		
Value of the risk behaviour indicator	—	RBI = 2%
Area covered by the BBS process	—	OI = 20%
Level of activity of qualified observers	—	AQOI = 80%

Source: own elaboration.

observations to be carried out during working hours, as production rates were the most important consideration for them. A fproblem reported by observers was also the low quality of feedback, due to a lack of skills and insufficient knowledge of the observed area station.

The motivation of the observers declined steadily after six months; the level of observer activity in AQOI dropped to 55%. Thus, the planned area under observation was only 15%. Obviously, there was also no improvement in accident statistics.

It should also be noted that, in general, the BBS process was well received by employees; a number of positive behaviours were observed, and employees responded well to positive feedback.

After seven months, the organisation decided to suspend the BBS process.

2.4.2 Why Was the Implementation Not Successful and What Mistakes Were Made?

- The first mistake was that an inexperienced person decided to do the implementation on his own, relying solely on the training materials. No implementation and enforcement teams were established, which should have included, in addition to the head of the H&S department, other representatives such as managers, staff representatives and trade union representatives. As a result, responsibility was not assigned to the members of the teams. While they may have been the same individuals, implementation requires the cooperation of many people with assigned roles. In the organisation in question, the attempt to implement BBS was essentially a solo effort by the head of H&S. Safety in an organisation is created by people at all levels; without their involvement, change is doomed to failure.
- Another mistake was inadequate communication. From the moment the decision was taken to implement it, all employees needed to be informed about the various stages of implementation and systematically updated on the effects of the BBS process.
- Managers were not informed of their role in the process, and tasks and responsibilities were not assigned to them. It is the managers' responsibility, firstly, to create conditions that enable the observers of the department concerned to carry out the planned observations. Secondly, the manager's role should be to monitor the observers' fulfilment of the planned number of observations, failing which they should initiate corrective action to meet the targets. Thirdly, managers should analyse the results of the observations carried out in their section or department in order to communicate these results to their subordinates and take possible corrective actions. In the company in question, managers were not involved in the BBS process at all.
- During the initial stage, no trial observations were carried out, during which it would have been possible to discuss the relevant points (risks). Unfortunately, in the organisation, no list of dangerous behaviours (potential risks) was constructed, which would have been especially for observers without knowledge of H&S in the observed area.

- Unfortunately, during the implementation and functioning of the BBS process, the observers were not provided with support or professional care, such as coaching and mentoring. Left to give feedback on their own, the observers found this situation to be demotivating.

In conclusion, the company's decision to suspend BBS seems premature. The effects of the BBS implementation should have been anticipated in the long term, and the lack of results after a few months of the process should have prompted corrective action. In the current situation, stimulating members of the organisation to take action and get involved will be a much more difficult challenge.

3 Change Management

How Change Becomes a Real Opportunity for Organisational Development

Rapidly occurring changes in various social and economic spheres, resulting from objective laws of development, affect the functioning of enterprises. They are also one of the main reasons for their success or failure. These processes are never indifferent to the well-being of employees, as they affect their sense of stability, as well as their safety, including safety at work, understood as the protection of health, life and psychophysical integrity in the environment of its performance. Successful implementation of change in an organisation, therefore, requires not only structural and procedural aspects to be taken into account, but also psychological factors of human perception and adaptation to changing conditions. Their omission can lead to lower levels of trust, increased tensions and a sense of insecurity among employees, which directly affects the level of safety culture in the organisation (Al-Haddad & Kotnour, 2015; Burnes, 2011; Errida & Lotfi, 2021; Jacobs et al., 2013).

Change management has become an essential issue for modern organisations. Increasing competition, globalisation, unrestricted flow and access to information have meant that only flexible and innovative organisations can survive and succeed today. However, in order to successfully plan and implement the desired changes, extensive knowledge and experience in change management are needed. This knowledge includes the psychological aspects of human behaviour in the face of environmental change. It cannot be underestimated, as it is employees who design change, are responsible for its implementation, adapt to it and learn new solutions. Their attitude towards change is the determining factor for the success or failure of the actions taken.

In order to manage change effectively, managers need to be equipped with knowledge of the various mechanisms, constraints and psychological obstacles affecting the effectiveness of the whole process. The theoretical output on change, contained in numerous studies, is very extensive. It will be presented in a condensed form in this chapter. It includes knowledge about the causes of change, types of change, areas of application and implementation models and their consequences. Organisational practice and new concepts of personnel management are increasingly updating existing theoretical constructs by transferring them to contemporary practical applications.

DOI: 10.1201/9781003677741-3

Good communication, employee participation and commitment building are effective ways to minimise resistance to change and build lasting commitment to it. Their application, combined with current approaches to organisational HR management and development, creates a kaleidoscope of possible implementations for effective change.

Change management takes on particular importance where job security is at stake. In this sphere, ineffectiveness or negligence may reflect not only in material losses, but also in the loss of health or life of employees (Dobrowolska et al., 2020).

The aim of the chapter is to highlight the importance of psychological factors in the process of effective change implementation and the role that employees at all organisational levels play in it. Practical ways of inducing people's commitment to the process and the real interactions conducive to overcoming the vapours against it will also be indicated, with a particular focus on the area of organisational job security. Common mistakes accompanying the implementation of change will be shown, as well as solutions from different companies that can help to avoid them.

3.1 UNDERSTANDING THE NEED FOR CHANGE AS A KEY TO INDUCING MOTIVATION

Changes in an organisation can be carried out to varying extents and frequencies. They may concern the management system as a whole or its components, for example, technical, social, economic and financial subsystems. They can also take on a multidimensional form; when implemented on numerous levels simultaneously, they result in a transformation of the company's organisational culture, including its safety culture. Moreover, they can span different levels – from the organisation as a whole to individual organisational positions. The aim of any change should always be to improve, to refine the organisation's existing way of working and, with regard to occupational safety, to optimise the conditions conducive to it. For this to happen, it becomes crucial to understand the subtleties of this complex process linking the numerous interactions within the organisation and externally, in its environment, as well as the need for understanding and acceptance by employees. To achieve this, management commitment should not only focus on the economic or technical scopes of change, but also take into account the delicate social factor.

Correct communication, participation in the planning and implementation of change with broad social groups, and understanding of resistance due to feelings of loss or threat among staff are often the key to the success of planned changes. From a technical point of view, change management is very similar to project management. Change in an organisation is treated as a special kind of new product, which is called a *new organisation*. As with any project, a project leader, team or organisational unit responsible for implementation is appointed, a budget is prepared, an action schedule is prepared, and so on. Nevertheless, change management is, from a broader point of view, a special kind of process.

This process significantly disrupts the prevailing status quo of the structure and people in an organisation (Koźmiński & Piotrowski, 2006). It is not indifferent to the pillars of an organisation's safety culture, as it affects the basic elements of organisational culture, such as values, norms, rituals and organisational roles, discussed in more detail in Chapter 1, disrupting previously existing interactions and patterns of behaviour. One of its key elements – safety culture – is not unaffected. Any change, whether it involves new technology, reorganisation of work or increasing cultural diversity among the workforce, interacts directly with existing safety practices and beliefs. It can weaken the existing pillars of that culture; however, if properly managed, it also provides an opportunity to strengthen them and give the safety culture a new quality. The change process should, therefore, be seen not just as an operational endeavour, but as a profound cultural intervention. Its overarching goal, in addition to achieving tangible organisational results, should also be to support and develop safety culture by consciously addressing its values, communicating expectations, modelling safe behaviour and involving employees in the change process. Only then is it possible to permanently embed change in organisational practice and build a working environment based on responsibility, trust and a shared system of meaning.

In the literature, we come across many definitions of the term *change*. On the one hand, change can be defined as any significant modification of some part in a distinguished whole (Griffin, 2005). Wendt (2010) states that change is a planned and controlled modification of an organisation's way of functioning as a reaction to anticipated or ongoing transformations in the environment, aimed at increasing the level of competitiveness and efficiency of that organisation. Kożuch (2011) emphasises that change is also a function of the system striving to improve and enhance its position in the environment, that is, a purposeful and conscious action aimed at moving from the current state to a different state. On the other hand, the scope and structure of change are determined by trends in the environment. Each of the aforementioned definitions is different; each represents a different approach but, in fact, each is intended to convey the same idea, which is best expressed in the words that change is a transition from the current state to a different (desired) state.

Planning and implementing change in an organisation can, according to Koźmiński and Piotrowski (2006), occur under several models. Their effectiveness depends on the scope and pace of implementation.

The directive model involves a very small group of people from the organisation in the planning and implementation of change, primarily managers and specialists from selected staff units, with line managers rarely included. Typical areas of change in this model are largely economic, technical or marketing procedures that are fundamental to the organisation. This approach is best applied when the company operates in a relatively stable or supportive environment and the scope of the change itself is not profound. The process of communicating the changes being introduced does not rely on lengthy communication channels when the company's structure is extensive; contact between managers and staff is direct. In the area of occupational safety, this may involve the implementation of new procedures for operating machinery/equipment or changes to the current circulation or format of documentation. An important element in this process is the provision of clear, unambiguous information to the workforce regarding the scope of action currently adopted, highlighting areas of change. This can be based on drafting and discussing instructions, as well as preparing a plan for the necessary training and its implementation.

The change transporter model is used when planning and implementing change in structures, personnel policies or information systems. It is believed that these types of changes will translate into other changes by creating organisational conditions for their implementation (e.g. future planned technological or organisational changes). A small group of employees is involved in the creation of these conditions, as in the directive model, but they consequently become leaders and promoters of change at subsequent stages. Changes of this type constitute a kind of transmission belt for future actions; for example, the development of elements of periodic evaluation of employees concerning their safe behaviour may form the foundation for a system of promotions in the organisation conditioned, among other things, by the observation of the employee's actions in this field.

The coalition-building model aims to jointly develop a concept for change and implement it collaboratively. The intention of this approach is to form a strong coalition for change, with agreed views on the company's mission and strategy. Managers, especially line managers, are expected to be communicators of change ideas as well as the individuals responsible for implementing them. With regard to occupational safety, it is important to identify participants with a high degree of social trust, who not only become promoters of change, but also model decent attitudes and behaviour.

The values management model is based on the long-term modelling of an organisational culture that fosters flexibility and change. It is an approach in which the entire workforce of the company is involved during the process of formulating and implementing change. Emphasis is placed on promoting and assimilating priority, commonly agreed values. The application of this approach generally requires a longer period of time, well-informed managers and employees about the direction and need for change, and the use of a rich set of techniques, mainly from the area of company social potential management. With regard to safety culture, we can speak here of the implementation of programmes that continuously promote the actions of the entire workforce in favour of safety through openness to report initiatives to improve working conditions or participation in the BBS process discussed in Chapter 2.

In conclusion, it is not possible to determine which of the models outlined earlier is best for a particular organisation. Any of them can be effective if the situation faced by a specific company is correctly assessed by its management. It is important to remember that the approaches presented here are typically conceptual models of change. In reality, practical solutions often incorporate elements from different approaches and are applied simultaneously in organisations.

Every organisation functions as a complex system in which two closely interrelated systems coexist: the technological system, comprising the requirements of task performance, the physical environment and the work tools used; and the social system, that is, the network of mutual relations and interactions between employees (Kożusznik, 2002). Both of these systems are in dynamic interaction; the way work is organised influences social relations and, in turn, the characteristics and behaviours of employees affect the way tasks are performed. In order for an organisation to function and develop under changing conditions, it must constantly adapt to its environment. This environment is not homogeneous; it is made up of many interrelated elements that interact both with each other and with internal organisational processes. This web of interdependencies shapes the context in which changes in the organisation take place.

In the literature, the individual components of an organisation are often referred to as elements or subsystems such as structure, technology, people, tasks or the external environment. In many change management concepts, particular importance is given to the human element, which is not only a resource, but also an active participant in transformation processes. Consideration of this aspect is crucial, especially in the context of changes affecting organisational culture and job security.

Likert (1967) considers organisations as a system of interconnected groups, where the link between groups is formed by individuals occupying key positions of "dual membership". These individuals are, so to speak, the link between the organisation and the environment. By virtue of occupying certain positions in both the organisation and specific systems of the external environment, they can become buffers of change.

Katz and Kahn (1979) take the view that an organisation can be considered through an arrangement of roles. The organisation as a whole can, thus, be seen as a collection of overlapping and interrelated roles – some of which may transcend the organisation. The plasticity of the roles and their variability affect the success of the changes introduced in the organisation.

Szczupaczyński (1995, 1998) considers the functioning of an organisation based on the four subsystems that make it up:

- people (their motivations, habits, attitudes, qualifications, skills, energy, learning abilities, interpersonal relations, etc.);
- tangible and intangible resources (buildings, machinery, information collected by the organisation, technology, capital, etc.);
- structures (formal and informal rules of interaction, division of labour, responsibilities, power, privileges, information, etc.);
- goals and objectives (strategic objectives of the organisation, operational objectives and tasks carried out at individual workplaces).

The organisation achieves its goals through integrated action at the level of all the subsystems that constitute it; a change in any one subsystem determines changes in all subsequent subsystems. The achievement of objectives is, in itself, a process. The researcher distinguishes four basic processes that take place in any organisation:

- the process of interaction between the organisation and its environment, which influences the organisation's goals and the possibility of achieving them (e.g. rapid changes in the credit market affect the activities of those enterprises that depend on external sources of financing; conversely, the enterprise can, with the help of product promotion tools, influence customer tastes and, thus, shape the elements of the environment that are important for the results of its activity; legislative changes concerning work safety force a cascade of changes in organisations, which – in turn – influence subsequent regulations of a legal nature);

- the process of setting and implementing successive organisational goals, whereby the organisation's goals are devised and planned by the management and then implemented through the joint efforts of the whole organisation (e.g. after achieving the planned economic results, the company's management modifies the development strategy and begins activity in new areas, which involves setting new tasks for the organisational units and the people employed in them, often including those related to occupational safety);
- the process of transforming subsystems of the organisation and their interactions (e.g. the development of a new domain of the organisation's activity may imply the need to purchase new equipment, apply new technologies, employ new staff, change organisational structures);
- the process of consumption and renewal of the organisation's resources (in order to exist, the enterprise needs to renew its machinery, renovate its buildings, train its employees, etc.).

One of the more well-known studies is by Leavitt (1964), who believes that changes in an organisation can be implemented by altering the organisation's structure, technology, people and work organisation. The change in the organisational structure of the company is most often the result of the organisation's adaptation to technological requirements, which entails a modification in the area of work organisation within the company. Thus, changes in each of the elements mentioned directly or indirectly affect the individuals employed in the organisation. The same author considers that changes in the structure of the organisation can take the following forms: classical, decentralisation and modification of the course of organisational processes. All of the areas of organisational change mentioned by Leavitt (1964) entail serious consequences, are time- and energy-consuming and can measurably threaten the security of the interests of individual employees or groups of employees.

Classical changes that result in a flattening of the organisational structure simultaneously lead to a reduction in the number of managerial positions within the organisation and, consequently, to the demotion of a large group of existing managers, team leaders and directors. Such actions also limit the possibility of vertical promotion within the organisation. The alternative becomes horizontal promotion – a specialist career – for which many restructured companies are unprepared. Internal regulations and procedures allow for the acquisition of privileges, raises or simply an increase in the scope of activities and responsibilities only through promotion to a managerial position.

Decentralisation, that is, the hiving off of subsidiaries from the company, often results in the loss of previously held privileges provided by the parent company to employees. In large organisations, there may be pressure from trade unions to provide social protection for employees of hived-off entities in order to maintain internal regulations concerning, for example, financial benefits such as jubilee payments, industry allowances and pre-retirement benefits. Spin-offs often extend some of the existing privileges for a limited period, but some are not secured by the new employer. Such a situation often results in a sense of disadvantage among employees of separated entities, and not infrequently this is also associated with their strong opposition.

Classical change should be understood as changes that are expressed in the creation or liquidation of units, changes in the span of management, competencies and responsibilities. Decentralisation, as a form of change, has a different purpose and course. The guiding thought here is the belief that smaller units create better stimulating conditions for efficient work, and in the process of achieving this, they themselves create the right structure and use the right technique.

The third form of structural change, that is, the modification of organisational processes, entails changes in existing working methods. A change in one section of work, in one department or division, will not bring the desired results for the entire organisation if the others do not also adapt their activities to the current requirements. Although a properly established course of organisational processes (workflow, decision-making process) improves the entire operation and contributes to productivity and job satisfaction, it can also cause concern among employees. These concerns relate to whether they will be able to cope with their new responsibilities – and those who know they have little time left before retirement may passively or actively avoid tasks that require the acquisition of new skills.

Given the many links that exist between the elements of an organisation, attention should also be paid to the diversity and variability within them. The subsystem most relevant to this book, namely the human subsystem, can change even when the company's workforce remains structurally and personally constant, regardless of the changes that adapt the organisation to its environment.

All of the changes characterised earlier are reflected in the workplace safety culture. Its dimensions, as part of the change, are subject to modification. If this process is not properly managed, it can lead to a disruption of previously effective communication channels, a lack of identification with the organisation and its goals, a loss of trust in authority or a reduction in motivation and commitment, resulting in a sense of loss of opportunities to have one's needs met in the future and a lack of consistency in norms of behaviour in an emergency situation.

Given the scopes of organisational change outlined, an effective planning and implementation process should include clarity on the following criteria:

a. the reason for the change (internal vs. external),
b. predictability of the change (planned – anticipates and prepares the organisation for expected future events vs. adaptive – responds to circumstances and emerging events),
c. depth of change (superficial – general and refers primarily to the need to adapt to changes in the environment vs. deep – strongly transforming the current state of affairs),
d. speed of change (rapid – requiring immediate response vs. slow),
e. the method of change (imposed – introduced top-down without taking into account the position of subordinate entities and without including, e.g., employee participation vs. agreed and accepted by all stakeholders),

f. influencing the environment (psychological – refers to the sphere of relations vs. economic – occurring on the grounds of the economic conditions of the organisation's functioning; technological – refers to the technological sphere (Sokołowska, 2009)).

In the case of organisational change, the process of planning it generally involves four main stages:

1. identification of strategic problems,
2. diagnosis of problems,
3. preparation of a change project,
4. recommendation of change implementation.

In many organisations, the technical and economic aspects of the planned change are considered at the levels mentioned above. However, these are only fragments of this complex process. Failure to take the human factor into account puts the change process at risk of failure, even at the stage of creating the framework for its implementation.

An interesting approach that provides an opportunity for a broader view of the change process in an organisation is the scheme proposed by Koźmiński and Piotrowski (2006) called the *transformation triangle*. The transformation triangle is an integrated approach to change, combining various separate initiatives into a coherent programme of action. The change paradigm here exposes rules of conduct such as:

1. Establishing a top-down direction. This activity consists of creating a concept for the entire organisation after the change and developing the conditions for putting it into practice (e.g. through elements of HR systems);

CASE 1

As part of a planned organisational change related to the launch of a new production line in a company in the construction industry, the recruitment of 150 executive employees is planned. In the area of activities related to internal promotions to leadership positions in the newly created department, changes are planned, preceding this process, in the area of the current creation of career paths in the organisation. The new measures within the promotion procedure include the condition that the candidate for the position of leader should have no accidents and that s/he should take initiatives to create a culture of work safety within the framework of the programmes implemented in the company. New rules for internal promotions have been developed and implemented throughout the organisation.

What elements of an employee's work safety activities should be included in promotion procedures?

2. Bottom-up problem-solving in the change process. In this rule of conduct, the idea is that people from all organisational levels should undertake and experience the problem-solving embedded in the change concept and participate in the implementation of the solved problems;

CASE 2

An organisation in the steel industry decided to develop a process for employee participation in the selection of personal protective equipment. The project involved the individual brigades selecting their delegate – the Shift Leader – who was given four types of protective eyewear to test. The tests lasted for three weeks; the leader's task was to discuss with the brigade the advantages and disadvantages of the tested glasses and identify the best ones. The organisation took the individual brigade's choices into account after the testing was completed. This process was the beginning of a wide-ranging change involving the systemic participation of the workforce in various activities to improve working conditions and safety.

In what areas of change in your organisation can the role of Change Leader be helpful?

3. The cross-functional process of redesigning an organisation's core business. The essence of this rule of thumb is to bring together different organisational activities and functions with the aim of improving their delivery;

CASE 3

A bottom-up selection of crew members enjoying the greatest trust and authority among employees allowed the launch of a "Leader of Change" programme in one manufacturing company in the construction sector. Individuals earning this designation were systematically involved in the organisation's change planning and implementation activities. They were also a very effective link in top-down communication. At the same time, performing this function was an important argument in the planning of the employee's development path and was taken into account during job promotions.

How can the social authority of staff members translate into the implementation of changes in the area of occupational safety?

The rules outlined above form a coherent whole and should not be considered components of separate activities. A holistic approach to change planning becomes the starting point for implementation processes within the above framework. The principles outlined should accompany the entire change planning and implementation process. This process can be divided into four main stages.

The first step is to establish the direction of the change. During this stage, based on an analysis of the organisation itself and the external environment, it becomes important to identify the most significant stakeholders in the process. Depending on the scope of the change, these can include individual positions, departments, divisions of the organisation, organisational units, management levels and even local government bodies and external institutions (Stasiła-Sieradzka et al., 2020). This identification is a key element in diagnosing the opportunities and threats that may arise during the change implementation phase. This activity also serves as a resource for building clear communication channels, allowing for an unambiguous message regarding the need for change and generating ideas to help overcome potential resistance. The diversity of the participant environment becomes an important aspect; it increases the ability to analyse the risks associated with the change from various perspectives and provides the opportunity to benefit from experiences outside the organisational environment and from other companies.

CASE 4

A high occupational safety culture is a differentiating factor for a company and influences its competitive advantage, but creating it requires the involvement of all employees. Continuous monitoring of the safety climate is becoming one of the important elements of this process. It is a kind of "barometer" of employee moods, knowledge of which provides an opportunity to model appropriate ways to improve safety. Any research that not only allows the assessment of the safety climate but also creates a field of comparison and the search for effective solutions for the introduction of appropriate changes becomes the basis for working out practical implementations and, as a result, the improvement of the comfort and safety of employees. Internal benchmarking is one of the methods that allow this to happen. Benchmarking studies on the dimensions of the work safety climate carried out within opencast and underground mines in Poland have identified numerous differences between these organisations. The joint development of a scope of change based on the best diagnosed solutions has significantly contributed to improving occupational safety (Stasiła-Sieradzka & Dobrowolska, 2016).

What areas (divisions, departments) in your organisation can be compared within the work safety climate diagnosis? How can the best solutions identified for creating safe work be implemented in further areas of the organisation?

The next stage is the design of the change process. This involves translating the vision for change into specific implementation goals. At this stage, it is good practice to create project teams whose task is to ensure clear communication during the implemented stages of change. In addition to communication, the team's responsibility is to provide support to line managers, who are often directly involved in the change implementation process and feel the consequences of pressures against them. Another task is establishing quantifiers for the changes being implemented, setting milestones and defining a process for monitoring the entire procedure. In addition to hard measures of change implementation, such as economic performance and the stage of technological redevelopment, it is important to establish measures of the workforce's mood. Methods such as focus group interviews, where staff are asked about their problems related to the successive stages of implemented changes, and systematic meetings with Change Leaders, who report on their bottom-up perceptions of the implemented stages of change, can be useful here. Such activities make it possible to monitor the team's motivation to participate in the change, forming the basis for counteracting resistance to the change (Appelbaum et al., 2017).

CASE 5

In a company belonging to the steel industry, a process of change implementation was planned in connection with the construction and commissioning of a new process line. The change was to involve the expansion of the plant, the internal allocation and training of some employees and a change in work organisation. In order to correctly implement the change process, a working team was set up, consisting of crew representatives drawn from the company's various departments, who enjoyed public confidence. They were divided into teams of two and trained in the methodology of conducting focus groups. The main task of the selected teams became to conduct focus groups on a monthly basis with crew representatives belonging to their departments. This included discussions based on predetermined theses, as well as on topics specific to a particular area of the organisation. The next step was to discuss and share the information obtained, which became a very valuable set of data allowing for the modification of the change process and appropriate adjustments at a very early stage. These activities provided an opportunity to involve the entire workforce in the change process. Employees have gained a sense that their voice matters, that they are being treated subjectively and that they are a valuable resource for the success of the change process.

What information, in terms of opportunities and threats, gained through the focus group method, can be useful for planning and implementing change in your organisation?

The next phase is the implementation of the change, that is, the realisation of the organisational redesign. In this phase, the "change triangle" is implemented in its entirety. Communication discipline and the use of integrated ways of managing the whole process are extremely important at this stage. The biggest problem of implementations is chaos: the lack of clear personal responsibility for the various stages of the process – established "milestones" – as well as the excessive diversification of tasks, leading to a lack of synergy between the various cells of the organisation. This can result in delays, a cascade of minor problems and omissions in action due to incompatibility.

The final stage is to institutionalise and root the change. To be effective, it often requires not only the implementation of new technologies or management practices, but also the creation of values, including those related to occupational safety.

CASE 6

Following a change in an organisation in the construction industry, which involved the implementation of innovative construction techniques using new lifting machinery, a programme for the exchange of information between individual brigades was implemented. Each concluding brigade filled in together a short protocol of the day, in which they summarised their observations, comments on the operation of the machines, the organisation of the work, as well as providing information on unusual or new events observed and solutions applied. This information was transcribed onto a graphic change map, which provided a visual representation of the process evaluation (the map was placed in a location visible to all). These activities sensitised everyone to constant observation of the workspace – especially in its dangerous areas. They also provided an opportunity to quickly eliminate perceived risks and modify the work organisation.

Does your company have a developed solution for the rapid communication of observed abnormal or hazardous events within the operation of new equipment or implemented work procedures?

3.2 THE PRINCIPLES FOR EFFECTIVE COMMUNICATION ABOUT CHANGE ACTIVITIES

Communication accompanying the change process is a key factor influencing the effectiveness of its implementation and the sustainability of its effects. In the context of shaping and strengthening a safety culture, the manner in which information about planned changes is communicated takes on particular importance, as it directly affects the level of trust, commitment and sense of security among employees.

Information about the change is disseminated through the various communication channels available in the organisation – from face-to-face conversations with employees and team meetings, to messages published on internal information systems (e.g. intranet) and formal instructions and communications addressed to managers (Smith et al., 2014). However, alongside formal channels, there are also informal sources of information that often create a distorted, fearful and distrustful picture of impending change. In the context of safety, this can result in uncertainty about future standards of health and life protection, liability rules or expected behaviour in emergency situations. This is why transparent, consistent and thoughtful communication about the change becomes an essential condition for its success and for maintaining a high level of safety culture. Employees should receive clear, understandable and regularly updated information – both about the objectives of the change and its impact on their daily work, responsibilities and safety procedures. Effective communication should also take into account the emotional dimension of the change; it is particularly important to recognise and address employees' concerns that may arise from uncertainty, a sense of threat or loss of control. Only then can a safety culture based on openness, trust and shared responsibility be built.

Communication plan. Creating a communication plan is an important element and should take place during the planning process of the change itself. One significant mistake is to prepare it only at the stage of implementation when, influenced by the rush of ongoing tasks, messages are communicated in a chaotic, unplanned and spontaneous way. Such actions can create the impression of a lack of control over the process itself, resulting in increased fear and anxiety. When creating a plan, it is necessary to determine the target group, that is, to specify to whom the information about the planned changes and their stages of implementation should reach, what channels of communication will be used, what frequency is planned for specific messages and who will be their sender (management board, managers or directors of individual organisational divisions).

Purpose of the change. In the process of communicating the planned change, a very important element of communication is the explanation of its purpose. The reason for the change must be clear, presented as relevant to the organisation, solving the diagnosed problems and bringing tangible benefits on many levels. A message that allows for an understanding of the real benefits, including the direct benefits to the individual employee receiving the message, is a key element in eliciting their commitment to it.

Tailoring the message to the recipient. Both the language of the message itself and its form and frequency are important. We probably have very different audiences in most companies, with the exception of those tightly homogenised small organisations, and the same message to the whole workforce may simply not be good enough. We will speak to everyone and no one at the same time. It is important to communicate the change as many times as necessary until we are sure that everyone has understood the purpose and meaning of the change.

Openness to bottom-up communications. As part of communication, it is useful to not only create a path from the top down, but also pave the opposite direction. There should be an opportunity to ask questions in face-to-face meetings or an email address to which queries can be sent. Even an anonymous box to collect the concerns of the crew provides the chance to hear concerns and interesting solutions from employees. This makes it possible to improve the messages silencing the concerns raised, as well as making employees participants in the whole process.

Information cascade. Leaders, both formal and informal (people given authority by the group), are often much more effective at convincing others of change than formal messages and have a greater influence on how the change is received. It is worth identifying such individuals, ensuring that they understand and accept the importance of the planned change and become advocates for it.

Information mirror. It is worth checking that the message we are conveying is received in the right way. The sender's intention is not always consistent with what the receiver hears and understands. It is important to give the listener the space to express how they are receiving the messages. This allows you to catch, at an early stage, potential distortions of the message. Gathering information about the information channels used is also very legitimate. It can happen that the planned form of communication turns out to be ineffective; for example, an email message may get lost in the crowd of messages received by an employee, information in the form of posters may be placed in the wrong location within the company's buildings or meetings may be scheduled at inconvenient times for the staff.

CASE 7

In one of the Polish branches of an international construction company, an obligation to wear protective glasses while working and moving around in a part of the production hall was introduced. The order was not preceded by information explaining the rationale for such a requirement; no such protection had been used since the factory's inception, and the technological process had not changed in the meantime. The injunction was met with resistance from the workers. They considered it overprotection, a nuisance at work and an apparent concern of management for their welfare. As a result, the health and safety services recorded numerous abandonments of the use of glasses.

Communicating to the workforce the reasons for the implemented change altered their perception; here are the elements of the message:

1. The demand for protective eyewear is a bottom-up initiative, having been made by employees of a company branch in another country;
2. The demand arose after an accident was recorded involving a worker's eyes being blurred by elements of post-production dust, which proved to be a serious hazard;
3. The incident resulted in the consequences of a long-term illness and significant loss of sight for the worker;
4. The research indicated that the proposed eyewear model significantly mitigates the potential hazard and is an important element in protecting workers' health.

As a result of a correctly conducted information campaign and showing the perspective of the implemented change in the area of occupational safety, the requirement to wear glasses was well received and accepted by the workers.

Do you see any possible areas for improvement in communication about changes in your organisation regarding occupational safety?

3.3 INVOLVING MANAGEMENT AND WORKERS

The involvement of both management and line workers plays a key role in the process of introducing and implementing change and is particularly significant in terms of building and embedding a safety culture within the organisation. Without the active participation of these groups, any changes – even the best designed ones at the formal level – remain mere declarations, not translated into actual attitudes and behaviours.

The basis for effective implementation of change, including in the area of occupational safety, is the universal principle of reinforcing commitment. Although they may seem simple, their practical implementation requires a comprehensive and systematic approach. These elements do not function independently; they form a coherent, mutually supportive system that fosters an organisational environment based on responsibility, trust and continuous improvement of safety standards.

Personalisation of benefits. The most effective element for building employee engagement at any level of management is the personalisation of change benefits. The needs of people realised through work are very diverse due to their career level, status within the organisation and other factors. However, for all, finding personal benefits in change is extremely important. One of the main mistakes organisations make is neglecting messages that emphasise this very principle in the communication process. It is important to construct these messages in such a way that the employee has the opportunity to identify the maximum amount of personal benefit. Even insignificant issues, such as easier and faster circulation of documents following the introduction of a new IT system, reducing the amount of printer paper ordered by a person or the withdrawal from use of unreliable office equipment that made daily tasks more difficult, can serve as a strong incentive to engage in the change process. It is also important to demonstrate the real, short-term and perceived benefits of the change, highlighting the advantages of each stage of change implemented (Kanter, 2011; Kotter, 1995; Mento et al., 2002).

In the context of creating a safe working environment, communication about planned changes should emphasise the specific benefits of caring for the health and well-being of employees. Particular emphasis should be placed on aspects such as improving the ergonomics of workstations, reducing exposure to harmful substances, eliminating activities that lead to excessive fatigue, and other solutions to support the daily functioning of staff in a safe and sustainable manner.

Participation. Building involvement by including wider staff groups in the planning and implementation of change is an important element in fostering staff engagement. This can be achieved through project workshops or focus groups composed of representatives from different employee groups, who collaborate to develop both the change plan and the stages of implementation. Collective decision-making, where possible, is a mechanism for sharing responsibility for the effects of implementation. Individuals participating in such teams become leaders of the change, often not only informing the change, but also able to highlight to other employees the real benefits of the change.

Supportive environment. Any change in an organisation involves simultaneous perceived opportunities and threats. Even the best planned and executed communication cannot completely eliminate the cognitive distortions that can generate anxiety. Employees' concerns often relate to potential losses, a lack of sufficient competence, limited resources or fatigue resulting from prolonged adaptation efforts. All of these factors can lead to a weakening of commitment, including the commitment to comply with and promote occupational safety standards.

In this context, building a supportive organisational environment becomes a key factor in cushioning the tensions associated with change. One of its cornerstones is open communication – based on mutual respect, empathy and a willingness to listen

– which allows employees to voice their concerns and receive regular, constructive feedback. It is worth referring here to the concept of relational crafting, understood as actively shaping workplace relationships by strengthening social bonds, seeking support and redefining interactions in a way that enhances a sense of meaning and psychological safety; this will be discussed in detail in the next chapter. Supportive relationships with supervisors and co-workers not only provide a source of information and emotional support, but also foster the internalisation of safety values, facilitating their rooting in everyday practices.

Thus, a work environment based on consciously built supportive relationships not only mitigates the negative effects of change, but also becomes a space conducive to the development of a safety culture based on trust, cooperation and mutual responsibility.

Modelling desired behaviours. The process of planning and implementing change can be long and aggravating. Building and sustaining commitment becomes easier if the whole process is divided into clear stages. However, these milestones must not only be based on technological or financial elements but also address the human side of the process. Identifying leaders who contribute to the success of a particular milestone, celebrating successes and sharing current experiences, becomes an important element.

3.4 OVERCOMING RESISTANCE TO CHANGE: THE ROLE OF SOCIAL AND ORGANISATIONAL SUPPORT

Despite initiatives to involve employees in the planning and implementation of change, resistance to change is a natural phenomenon. The anatomy of negative reactions – behaviours that indicate the existence of a motivation that prevents an individual from deciding to adopt an innovation or even compels its rejection – needs to be viewed from three research perspectives:

1. Individual – then we will analyse resistance to change using psychological concepts and methods,
2. Organisational – then we will be interested primarily in factors of the organisational environment as determinants of group resistance,
3. Resulting from the interaction of factors considered according to the aforementioned points of view (Ratajczak, 1980).

The magnitude and pace of change in the environment, and especially its suddenness, gives rise to the phenomenon known as strategic surprise (Ansoff, 1985). Its indicators are the growing difficulties in the functioning of the enterprise and the increase in the number of its elements that are increasingly unadapted to the new conditions, as the organisation's ability to act effectively is not sufficient to cross the threshold of the required adaptation. Resistance to change, however, is not only in the organisation itself; it is extremely important for its members to be convinced of the need for change, taking into account the fact that individuals may be unaware of the changes taking place in the organisation's environment and, thus, fail to see the need for change in the organisation itself.

TABLE 3.1
Organisational Sources of Resistance to Change

Sources of Resistance	Causes of Resistance
Organisational inertia	Inbuilt stabilising mechanisms based on various types of formal procedures for dealing with diverse areas of functioning, e.g., staff selection and selection procedures, training dealing with accidents or breakdowns slow down the introduction of change.
Systemic nature of functioning	The interaction of interdependent organisational subsystems entails the need to introduce changes in each of them, the need for precise synchronisation of activities. The size of the organisation and the number of interdependent sub-systems are directly proportional to the resulting pore towards change.
Group inertia	Group norms have an inhibitory effect on individuals interested in change.
Power structure	The more codified the division of roles in an organisation, the greater the resistance to change. Hierarchy and status differences between individuals create fears of losing their own influence and privileges; as a result, the structure becomes ossified.
Informational exclusivity	Fear of losing exclusive possession of relevant information among those in specialist positions.
Resource structure	Fear of depletion in terms of established resource allocation including, e.g., operating budgets, office premises, equipment number of FTEs.
Inertia of individual employee subgroups	Results from the diverging interests of the many subgroups that make up the company's organisational structure – here job divisions, age divisions, divisions related to organisational unit membership and so on.

Source: Own elaboration

We can list several reasons for pores against change. Some of these are anchored in the organisation, while others can be described as social (Table 3.1).

Change is always associated with a cost that the company has to incur in relation to the reconstruction of its own organisational structures and human potential. A prerequisite for change to occur is that the forces that support it outweigh the forces that resist it; balance in this respect is a guarantee of no change in the organisation. Clarke (1997) presents this relationship in the form of the equation of change:

$$(A+B+C)>X \tag{3.1}$$

Where:

A – dissatisfaction with the status quo
B – desirability of change
C – feasibility of practical change
X – The cost of change

The emergence of resistance among employees depends on how much employees have invested in old ways of doing things, as well as being linked to their reluctance to leave behind what is familiar (Clarke, 1997).

Penc (2000) outlines multiple reasons for resistance to change that relate to the social dimension of change.

The reasons for resistance include:

- a lack of understanding of the need for change or its meaning as a result of a lack of reliable information from an authoritative source;
- fear that the proposed changes will lower the existing professional or social prestige of employees;
- the belief that the change will create demands that employees will not be able to meet;
- fear that supporting the change will expose other employees to losses and expose themselves to those who oppose the change;
- Tendency to stick with old solutions, lack of conviction in new ones and unwillingness to take risks;
- little trust in superiors, lack of faith that their assurances will be realised and any gains will compensate for the old;
- fear of a deterioration of the relationship between the effort put into the work and the potential rewards;
- a struggle to "save face", support for change may mean admitting to previously made wrong decisions;
- conformist attitudes, reluctance to ask uncomfortable questions, to clarify doubts about the change;
- Group process dynamics, adopting majority views;
- belief in inappropriate change procedures causing employees to focus on defensive actions, blocking the implementation of change.

This researcher also cites various explanations for the causes of resistance that may affect the process of change introduced in an organisation. These are:

- the law of spite, the reluctance to tear down the old, established order and the tendency to return to the old state;
- habitual inertia, behavioural inertia and the desire to restore a disturbed balance;
- cognitive dissonance manifested by avoidance of information and situations incompatible with one's own beliefs;
- ethos of sticking with one's own, blocking new solutions due to incompatibility with one's own experience and uncertainty of the future situation;
- groupthink syndrome, relying on the opinion of the group, succumbing to environmental pressure;
- illusion of control, being in control of the situation;
- being satisfied with limited rationality, settling for "good enough", considering what is good enough;
- concentration on what is certain, not on what is possible;
- Anxiety restraint, fear of responsibility, chaos, loss, exposing someone to annoyance;
- Lack of a sense of common interest, disbelief in the good intentions of superiors;

- A sense of low self-worth, lack of self-reliance, lack of confidence in one's own abilities, lack of courage, unwillingness to take risks;
- Defensive self-consciousness, creating a positive self-image, fear of losing a good image, prestige;
- living in a state of defensive alertness, the threat of losing one's position and the ability to satisfy felt needs;
- defensive setting of motivation, excess of negative reinforcement, guided by motivation to avoid failure;
- A decrease in adaptability, a decrease in creativity, a decrease in sensitivity to positive stimulation from the organisation.

Outbreaks of social resistance in an organisation can be considered not only in terms of individual employees, but also taking into account the interests of the professional groups employed in the organisation. Clarke (1997) emphasises that they tend to arise first where there is a questioning of established patterns of action, challenging old ideologies. The researcher draws attention to the importance of trade union members and their influence on people working in positions at the bottom of the organisational hierarchy and often in danger of losing their jobs, middle management often not included in the change process itself and restricted in their professional development, and decision-makers – top management in danger of losing their existing influence and power. In this prism, resistance to change is a ground that not only slows down or even prevents the implementation of change in the organisation, but also creates an environment conducive to the degradation of existing norms of behaviour, including those concerning safety rules. Sabotaging change then becomes a dangerous phenomenon, translating into less attentiveness to possible risks.

The issue of resistance to change is also important in the context of the organisational culture of the company undergoing change. Resistance occurs both in companies where employees do not feel very secure due to frequent hiring and firing, and in organisations where employees feel all too secure due to job stability. Many have many years of service with the company and there is no influx of new staff. In the first case, there is a precautionary behaviour, a low willingness to take actions that involve the risk of making a mistake, and in the second case, a lack of a sense of threat and attachment to existing work rules and social relations in the company.

A change in the organisation also means a change in the previously respected terms of the psychological contract existing between the organisation and the employee. As a result, people fear that these conditions will be less favourable to them than they currently are. The situation of change creates anxiety for the employee on the one hand, and hope for improved working conditions or more interesting tasks on the other. Anxiety and hope are subjective phenomena that are internal regulators of human behaviour. They are mechanisms of emotional regulation. In a situation of change, these mechanisms are activated due to the fact that the situation can be both a source of danger and a source of benefit for the person. Anxiety, on the one hand, may be a reaction to the threat of having one's needs met – for example, fear of a reduction in salary, of being fired, of losing valued social ties, of having to improve oneself, of acquiring new competences, of being confronted with new demands to be met. Hope, on the other hand, may be aroused by perceiving the chance to learn new things, the possibility of professional development, more interesting tasks, a more secure working environment and so on.

Fears and hopes are, therefore, not two extremes of the same continuum, but different psychological dimensions, "pulsating" in a situation of change and directing the individual's perception towards the threats on the one hand and the challenges of the situation on the other (Kożusznik, 2002). Emerging fears, which may take the form of resistance to change, can be effectively minimised by monitoring the change process, as well as systemic solutions introduced in organisations as a permanent element of programmes aimed at HR management.

3.5 MONITORING AND EVALUATING THE EFFECTIVENESS OF CHANGE IMPLEMENTATION

Monitoring the change process is an ongoing and multidimensional process. Technological or financial milestones are measurable and easy to identify. The social factor, measured in terms of staff commitment to change, is not so clear-cut. It is very difficult to introduce a yardstick in this area, especially as the initial phase of the change planning and implementation process is fraught with numerous disruptions that are impossible to predict. The length, intensity and volatility of the whole process often result in fatigue and even burnout of the most committed employees. Sometimes a change planned and introduced successfully becomes difficult to anchor in the organisation, in the area of many behaviours old habits return over time, innovative tools or work techniques are displaced by previous methods of operation (Smith et al., 2014; Znajmiecka, et al., 2022). The introduction of change and its anchoring in an organisation are fraught with some of the most common mistakes.

One of the most common mistakes in the change management process is insufficient communication. The lack of clear communication regarding the goals and benefits of the transformation leads to uncertainty and resistance on the part of employees. Another obstacle is insufficient preparation, as well as lack of adequate training for employees. Ignoring the need for skills development in the context of change, such as the implementation of a new IT system or new technological processes requiring familiarisation with previously unfamiliar machines and tools, results in resistance and difficulties in adaptation. Another common mistake is the fragmentation of activities in individual areas of change implementation (various departments and divisions) without the commonality and synergy of the whole process. This results in the indifference of some managers to problems in other areas of the organisation. An effective solution is to actively involve stakeholders at each stage of the process, which supports the building of coalitions of support and ensures consistency of action. Failure to monitor and evaluate progress is a common mistake that prevents early detection of problems. Ignoring performance indicators at the technological or economic level and failing to regularly assess employee satisfaction lead to the continuation of ineffective actions. To prevent this, managers should introduce systematic progress monitoring and nurture feedback from the team.

The mistakes described earlier can be minimised on the basis of new concepts of HR management. They point to the increasing role of empowerment and co-responsibility of employees in relation to the successful implementation of change. The concept of job crafting, that is, the proactive and conscious shaping of work by employees themselves, which will be presented in Chapter 4, is reflected in many areas of effective change management. Previously developed concepts (Penc, 2000) find room for updated practical implementations in its prism and are constantly being refined in the grounds of different organisations. Their versatility constantly clashes with the diversity of organisational environments, providing opportunities to develop more and more interesting and effective solutions. The points presented in the following list are only a general direction for the introduction of structural and multifaceted solutions in the field of HR management, which can be helpful in the successful implementation of change in an organisation.

1. A constant search for innovative solutions (constant adaptation to a continuous commitment to small changes and improvements). This can be done by introducing systemic programmes conducive to encouraging employee suggestions for improvement, for example, the *Report an Innovation* programme, which collects, implements and rewards suggestions for workplace improvements from both individuals and teams of employees. Another way is to take care to elicit information from new employees on observed possible areas of innovation implementation in, for example, an *Install Yourself* programme focused on encouraging new team members to share their observations and implement them in an organisational culture open to innovation.
2. Creating a repository of information about the problems and process solutions surrounding the changes being implemented. This can take the form of an internal information space with free access for employees. Sharing information on solutions developed is an area for building internal support in the ongoing process.
3. Broad communication of problems that need to be solved (especially when there is an opportunity to generate beneficial ideas and, as a result, develop projects that address multiple areas of the organisation). This can be done by building project teams that bring together staff members representing different areas of the organisation and organisational levels. This can be done on the basis of the *Coalition for Change* programme, which aims to bring together and systematically work with the crew representatives most involved in the processes of planning, implementing and monitoring change at different levels of the organisation.
4. Shaping innovation-friendly organisational structures by removing complex control systems and freeing the organisation from unnecessary bureaucracy. This can be realised by introducing special communication paths during the change implementation process of the so-called "red line of communication". A short information flow path between the management level and employees allows the barriers accompanying the change process to be quickly spotted and removed.
5. Increasing the sense of empowerment by genuinely involving employees in the process of planning and implementing change. For it to be effective, it is necessary for everyone to be active and take responsibility for the course and results of work. This can be done by showing employees real influence

and impact on the changes and innovations being introduced. This can be done through systemic solutions in the area of HR management, for example, through programmes such as the "*Change* Leader", which promotes the most active employee attitudes towards changes in the organisation and the provision of support.

6. Continuous attention to building an organisational climate conducive to innovation, based on openness to change and promoting readiness to take on challenges. The implementation of this assumption is multidimensional, but mainly based on a tolerance for mistakes and a readiness to analyse them in order to develop the most optimal solutions.

4 Safety in Your Hands

Job Crafting as an Activity Supporting the Building of Employee Involvement in Shaping Safety Culture

The previous chapters of this book have highlighted that effective management of workplace safety goes far beyond formal procedures, supervisory systems or engineering safety measures. An analysis of the BBS concept and a review of the determinants of safety culture indicated that it is employee behaviour, involvement, awareness and level of responsibility that form the foundation of a sustainable and effective safety system. Key importance was attributed to building organisational conditions that foster not only compliance with standards, but also active participation in hazard identification and continuous improvement processes.

Today's organisations operate in an environment of constant change – technological, economic and social – which raises new challenges for safety management. Increasing digitalisation, automation of processes, changing employment structures and evolving employee expectations mean that traditional management models – based on rigid procedures, hierarchy and control – are increasingly proving inadequate. The chapter on change management points out that flexibility, communication and the ability of organisations to involve people in changes that directly affect them play a key role in transformation processes. In this context, approaches that foster a sense of empowerment and co-responsibility among employees are becoming increasingly valuable.

In response to these needs, an interesting and increasingly considered proposal is the concept of job crafting, that is, the conscious, bottom-up shaping of work by employees themselves (Wrzesniewski & Dutton, 2001). Unlike traditional models, where responsibilities are strictly defined and imposed from above, job crafting is based on the belief that employees can – and should – actively influence the tasks they perform (task crafting), the individuals with whom they interact (relational crafting) and their perception of the meaning and significance of their work (cognitive crafting). This approach enables a better match between work and individual competencies, values and needs, fostering greater engagement, satisfaction and a sense of meaning – including in terms of security.

The approach assumes that individuals are not just passive performers of the tasks given to them but actively participate in the creation of their job role. Job crafting involves modifications in three areas: tasks (task crafting), work relationships

DOI: 10.1201/9781003677741-4

(relational crafting) and the cognitive perception of work (cognitive crafting). This enables employees to better match their work to their own competencies, values and needs, fostering an increase in motivation, commitment and a sense of meaning in the activities performed.

With regard to safety culture, job crafting can be an important complement to previously discussed strategies such as BBS. Employees who have a real say in their work environment are more vigilant, more likely to share observations and actively participate in prevention activities. In this way, safety ceases to be a mere obligation imposed by procedures and becomes a shared value and an element of organisational identity.

Importantly, job crafting can also act as an adaptation mechanism in organisational change processes. By enabling employees to reinterpret their professional role and redefine their professional relationships, it supports coping with the uncertainty and tension that often accompany transformations. In doing so, it integrates a security management perspective with change management practices, creating a coherent approach to building flexible, resilient and engaged teams.

The aim of this chapter is to present job crafting as a tool to support the development of safety culture in organisations and its potential in the context of change management. The first part of the chapter will discuss the theoretical underpinnings of the approach and its place in the broader landscape of research on employee motivation, well-being and engagement. This will be followed by a presentation of the practical applicability of job crafting in high-risk environments, taking into account the specifics of workplaces and safety requirements. The final part of the chapter will analyse the potential challenges, barriers and conditions for the effective implementation of job crafting, at both an individual and systemic level.

4.1 INTRODUCTION TO THE CONCEPT OF JOB CRAFTING

Job crafting, that is, the process of employees consciously shaping their jobs, is an important part of modern HR management strategies. In the context of a safety culture, job crafting takes on particular importance as it enables employees to actively participate in creating and maintaining a safe working environment. A safety culture, defined by shared values, attitudes and behavioural norms relating to health and safety at work, can be significantly strengthened through bottom-up actions – by workers themselves – in job crafting.

As demonstrated in the previous chapters, effective prevention is based not only on formal procedures, but – above all – also on the daily decisions, vigilance and responsibility of workers. Similarly, from a change management perspective, a real and sustainable change in organisational culture is only possible if it is based on the commitment of those affected by the change. Job crafting addresses both challenges; by linking individual needs and motivations to organisational goals, it enables a flexible response to changing working conditions while fostering a mature, proactive safety culture.

By being able to modify responsibilities, working relationships and perceptions of work, employees can better align their working environment with their own competences, values and expectations. Such activities not only increase satisfaction and commitment but also lead to higher levels of responsibility for safety – both their own and their colleagues'. Job crafting, thus, becomes not only a tool for personal development, but also a mechanism for strengthening organisational safety.

The concept of *job crafting* was introduced to psychology in 2001 by American researchers Wrzesniewski and Dutton. Currently, there is no single, universally accepted definition of job crafting. It is most often described as an employee taking action on their own initiative (often without the knowledge of a supervisor) to transform their job (Kasprzak et al., 2017; Minda & Kasprzak, 2018; Rogala & Cieślak, 2019).

In Wrzesniewski and Dutton's (2001) initial conceptualisation, the purpose of job crafting was primarily to make work more meaningful and relevant and to increase satisfaction, engagement, resilience and productivity. As research on job crafting has evolved, its definition has changed, and a more recent perspective on job transformation also includes adapting the way tasks are performed to an employee's individual preferences, abilities, motives and passions (Tims et al., 2012). Numerous studies, including meta-analyses (Rudolph et al., 2017), have shown many beneficial consequences of job crafting for both employees and organisations, such as increased work engagement (Frederick & VanderWeele, 2020; Tims et al., 2012), reduced boredom (Oprea et al., 2019), enhanced psychological resources (Rudolph et al., 2017) and well-being (Oprea et al., 2019). In addition, job crafting increases employee productivity (Dubbelt et al., 2019; Tims et al., 2012), creativity (Hu et al., 2020) and levels of citizenship behaviour (Kasprzak et al., 2017), which also benefits the organisation.

The first model of work crafting was developed by Wrzesniewski and Dutton (2001), who defined it as a change in the boundaries of work related to task performance, relationships with others and the meaning of work. The authors distinguished three main areas of job crafting: task crafting, relational crafting and cognitive crafting – the crafting of thinking about work (Kasprzak et al., 2017; Tims et al., 2012).

Task crafting involves modifying responsibilities, for example, by adding tasks that are more engaging, developing or in line with interests – or even giving new meaning to existing activities. In practice, this can mean, for example, becoming involved in the development of job instructions, analysing near misses or taking on a mentoring role for new employees.

Relational crafting involves actively shaping working relationships – both by avoiding aggravating relationships and by deepening those that foster cooperation, support and trust. In high-risk environments, this can take the form of creating informal safety groups, strengthening partnerships in work pairs or developing interdepartmental relationships to better coordinate activities.

Cognitive crafting is a change in the way work is perceived – seeing it as part of a larger whole, aligned with the worker's values and identity, making it more meaningful and relevant. For example, a machine operator may begin to see their job not just as "producing details", but as a key contributor to ensuring the safety of the final product or the continuity of the process. The crafting of work from this perspective usually takes place in three stages, starting with the motivation for change, progressing through the recognition of its possibilities and ending with the active implementation of change by the employee (Berg et al., 2021; Wrzesniewski & Dutton, 2001). Motivation to implement change may be driven by a desire to increase one's effectiveness in coping with work demands, to enhance one's sense of control over one's work, to improve one's self-image as an employee and to improve better interpersonal relationships with colleagues and superiors (Kapica & Baka, 2021).

Although Wrzesniewski and Dutton's model is popular from an academic perspective, it has also received criticism from some researchers. Leana et al. (2009) note that this model strongly emphasises the role of the individual while neglecting the importance of the team. In response, a distinction has been proposed between individual job crafting and team (collaborative) job crafting, understood as the joint actions of pairs or groups of employees leading to the modification of tasks, relationships or ways of working. This perspective fits perfectly with the assumptions discussed earlier in the chapters on BBS and change management. The BBS approach, based on observation, feedback and active participation of employees, assumes that safe behaviour is reinforced in a social context. Team job crafting can complement this process by creating a space for shared learning, mutual support and collective responsibility for safety. Furthermore, linking job crafting to the requirements of the ISO 45001 standard (International Organisation for Standardization, 2018) further reinforces its importance. Among other things, this standard emphasises the need for leadership and commitment, employee consultation and participation, communication and continuous improvement; such a link is presented in Table 4.1. Both individual and team forms of job crafting meet these requirements,

TABLE 4.1
Comparison between the Requirements of ISO 45001 and Job Crafting

ISO 45001 Requirements	Requirements of the Standard	Link to Job Crafting
Point 5.1 – Leadership and commitment	Creating an environment which supports active employee participation	Job crafting as a form of bottom-up shaping of working conditions – employees "craft" safety, ergonomics and sense of work.
Point 5.4 – Consultation and participation of employees	Consideration of employee opinions in decision-making	Employees have a say in the scope of tasks, relationships and work organisation.
Point 6.2 – OSH objectives and action planning	Setting health and safety objectives and involving workers in planning	Workers can propose organisational changes through job crafting, influencing safety.
Point 7.2 – Competencies	Ensuring the development of competences necessary to work safely	Job crafting supports the development of soft competencies (proactivity, feedback), relevant to safety culture.
Point 7.4 – Communication	Ensuring effective two-way communication	Open communication and feedback are key to both BBS and job crafting – enabling employees to initiate change.
Point 10.3 – Continuous improvement	Fostering a culture of improvement in the health and safety management system	Job crafting is a continuous, bottom-up improvement mechanism – employees identify and implement changes.

Source: Own elaboration

enabling employees to have a real say in shaping their work environment and to actively participate in improvement activities. Job crafting can, therefore, be seen as a bottom-up mechanism in line with the systemic approach to OSH management promoted by ISO – reinforcing a culture of involvement, shared responsibility and open communication.

It is worth emphasising that job crafting – whatever its form – is not incidental but is a continuous and dynamic adaptive process (Berg et al., 2021; Kasprzak et al., 2017). It can be undertaken by employees at different levels and in various industries (Slemp & Vella-Brodrick, 2013), and its effectiveness depends on both organisational factors – such as autonomy and innovation climate (Xu et al., 2019) – and individual characteristics, such as self-efficacy or proactive personality (Roczniewska et al., 2020). In the context of the previously discussed safety culture elements and the ISO-compliant systemic approach, job crafting appears as a practical and compliant tool to support employee engagement, competence development and continuous improvement of the work environment.

4.2 THE RELATIONSHIP BETWEEN JOB CRAFTING AND EMPLOYEE INVOLVEMENT

Contemporary approaches to occupational safety management increasingly emphasise that it is not only the formal implementation of procedures and standards that plays a key role in the successful implementation of an OSH policy, but above all the personal involvement of employees and their beliefs and attitudes towards risks (Boczkowska et al., 2022). An organisation is not only a system and a structure, but first and foremost people – their willingness to act, their awareness of risks and their sense of influence over their working environment. Effectively shaping a safety culture requires considering the specific cultural and environmental contexts and the diverse motivations of employees (Stasiła-Sieradzka & Znajmiecka-Sikora, 2017). Although formal measures such as the implementation of standards (e.g. ISO 45001, ISO 26000) are the foundation of a systemic approach to OSH, without internal employee involvement they often remain ineffective. Safety culture – as a set of values, beliefs and safety practices – cannot be built on documents alone. Its strength lies in people's everyday behaviour, which is driven by subjective risk perception, motivation and belief in the meaningfulness of actions taken. In this context, job crafting can be an effective, bottom-up tool to support the development of a safety culture. By making task modifications that enhance safety, developing relationships that support cooperation and trust, or reinterpreting risks in a more constructive way, employees actively co-create a work environment that matches their values and competences (Berg et al., 2021). As a result, their sense of agency, belonging and willingness to take action to protect both themselves and others grows. Organisations aware of their social responsibility should not only create safe working but also support the development of employees as co-authors of this reality (Boczkowska et al., 2022). This is because commitment does not mean merely following procedures, but stems from a deeper sense of meaning, identification with the organisation's goals and belief in the possibility of influencing one's own environment.

In the process of intensifying the creative and innovative behaviour undertaken by employees, it is crucial to create a work environment that fosters the exchange of ideas and autonomy, leading to greater employee engagement and participation (Hunter et al., 2007). The literature most often discusses different levels of worker involvement in decision-making processes for shaping safe working conditions, covering a continuum of the following activities: informing, consulting, co-determining and empowering, that is, delegating. It is also possible to distinguish opinion-forming as a more advanced form of consultation (Boczkowska et al., 2022). Employee participation in the area of occupational safety is related to the involvement of employees in the development of internal safety standards and documents, participation in initiatives and decisions and the introduction of a system of employee consultation on safety and other issues (Milczarek, 2002). In order to achieve employee commitment and dedication, it is necessary to create conditions that enable employees to meet their own needs and realise their potential. Commitment, a sense of meaningful work and proactive behaviours and attitudes related to keeping oneself and others safe, which are crucial for both employees and employers, result from autonomous action. Creating a work environment that promotes autonomy allows employees to feel more self-fulfilled and enhances their intrinsic motivation. Autonomous action, where employees have the opportunity to shape their tasks and influence their working conditions, leads to higher levels of engagement and an increased sense of responsibility for safety. Thus, the pursuit of employee engagement and dedication requires the implementation of policies and practices that not only support their autonomy and enable them to make autonomous decisions about their work, but also foster the desired beliefs and attitudes that precede behaviour. Hence, the range of activities that organisations initiate should translate into, among other things (Studenski, 1999, 2000):

- A belief in the value of complying with safety regulations – there is a strong belief within the company that complying with safety regulations and protecting the health and lives of employees benefit both the company and the employees. Employees are aware that these measures lead to a reduction in the number of accidents and occupational illnesses, translate into greater work efficiency and result in a reduction in costs related to employee absenteeism and medical treatment.
- An optimistic approach to improving safety conditions – an optimistic, open-minded approach is developed in the organisation, which assumes that it is possible to continuously improve safety conditions at work. Employees and management take action in line with defined objectives and standards, aiming to create a safe working environment in which the risk of accidents and hazards is minimised.
- Viewing life and health as key values – the company considers the life and health of employees to be its most valuable assets. Protecting these values is recognised as one of the main objectives of the business, on which financial resources and HR are worthy of investment. The company allocates resources to training, protective equipment and other activities that enhance employee safety. Equally, employees prioritise these values and put their own safety and that of their colleagues first in the actions they take.

- Management employees' responsibility for safety – managers (and/or supervisors) feel a responsibility to ensure safe working conditions for their subordinates. Their role includes not only supervising compliance with safety procedures but also being actively involved in identifying and eliminating hazards in the workplace.
- Observation and measurement procedures for compliance with safety standards – the company uses observation and measurement procedures to monitor compliance of conduct with accepted safety standards. Where unsafe behaviour or conditions are detected (usually by employees), immediate corrective action is taken. Regular audits and inspections ensure that safety standards are adhered to at every level of the organisation. The company fosters a positive attitude among employees towards audits as a form of safety enhancement rather than fault finding and enumeration.
- A belief in the value of investigating incidents, including those with no injuries, is developing within the organisation. There is an emerging consensus that investigating non-injury incidents, which could potentially lead to accidents, and formulating preventive programmes based on these investigations are as important as examining actual accidents and disasters. This preventive attitude fosters a better understanding of risks and helps to avoid serious incidents in the future.
- Removal of deviations from safety standards – an important element of a safety culture is the ongoing removal of any identified deviations from working conditions or the condition of equipment that do not meet applicable safety standards. A prompt response to these non-conformities minimises the risk of accidents occurring and is made possible by the initiative and attentiveness of employees and their approach to reporting such situations – not as "whistleblowing", but as a concern for the safety of colleagues.
- Willingness to inform management of the state of working conditions – regular preparation and communication of up-to-date information to management on the state of working conditions and their compliance with applicable safety standards is key. Such communication ensures that management is aware of potential risks and can take appropriate remedial action.
- Involvement of top management in safety culture – in order to raise the level of safety culture in the company, it is necessary to involve top management. They should actively participate in the development and implementation of safety policies and set a personal example in complying with health and safety regulations. Such attitudes and beliefs can lead to building co-responsibility and organisational togetherness and influence employees' willingness to undertake and promote safe behaviour.
- Formulation of preventive objectives – the enterprise should formulate preventive objectives and convince workers of their importance and take action to maximise their acceptance and achievement. These objectives should be clearly defined and communicated in a way that motivates workers to comply with them.
- Rational criticism – there should be rational criticism in the organisation that allows safety concerns and suggestions to be openly expressed. Such a

climate encourages the identification of potential hazards and the introduction of effective countermeasures.

- Promoting safety at work – an important element of a safety culture is to promote safety at work and to stimulate employee involvement in activities that protect life and health. This can be done through regular talks, training, posters, slogans and organising safety actions.

It is worth noting that a number of the areas described earlier are related to the individual worker's perspective and perception. If, at the level of thoughts and beliefs among working people, there is no willingness to engage in safe behaviour, then even the most committed actions on the part of the employer may not have an effect. From this perspective, it is worth extending the elements of employee participation to include activities that promote and support the crafting of work by employees. Employee participation and the crafting of work are two different but related approaches to employee involvement in shaping the work environment and the tasks performed (cf. Table 4.2). Worker participation, on the one hand, refers to

TABLE 4.2
Worker Participation and Job Crafting

Worker Participation	Crafting of Work
Purpose	
The aim is to involve employees in decision-making processes, which can lead to better management, greater acceptance of decisions and improved working conditions. It is about influencing whole organisational processes and policies.	It focuses on tailoring work to the individual employee's needs and preferences, which can lead to greater engagement, satisfaction and job performance. The aim is to make the work better suited to the individual.
Process	
The process is often initiated by the organisation, may be formalised by procedures/policies and may involve consultation with various stakeholders in the company.	It is an initiative of the employee, often without formal agreement with superiors, and involves the employee making changes in his or her own area of work.
Examples of activities	
Organising team meetings where employees can express their views on company policies, employee participation in supervisory boards and conducting opinion surveys among employees on working conditions.	An employee decides to add new tasks to his or her job description, e.g., additional monitoring of a procedure, changes the way daily tasks are performed, builds new relationships with colleagues or changes their approach to their duties.
The role of the organisation	
The organisation plays an active role in initiating and supporting employee participation by creating appropriate structures and procedures.	The organisation can support job crafting by creating the right working environment (e.g. giving more autonomy), but the initiative lies with the employee.

Source: own elaboration.

the involvement of employees in decision-making processes at different levels of the organisation. It is usually formal and organised, often driven by company policy or legislation (Milczarek, 2002). Job crafting, on the other hand, refers to actions taken by employees on their own initiative to transform their jobs. Job crafting is more individual and informal, focusing on the employee's personal preferences and needs. By equipping workers with the knowledge as well as the tools for job crafting, they can consciously, committedly and responsibly contribute to the safety culture of the organisation in which they work.

Employees are a key element of an organisation, and their involvement in shaping working conditions is essential for building an effective safety culture. Job crafting activities – although often undertaken intuitively – stem from employees' natural need to improve their daily work (Wrzesniewski & Dutton, 2001). Job crafting allows people to customise tasks, relationships and perceptions of their work, which increases commitment and responsibility for outcomes, including safety. Wrzesniewski and Dutton (2001) identify three basic needs that motivate job crafting: the need for personal control, building a positive self-image and establishing relationships with others. The need for control – linked to a sense of influence – is fundamental to mental health and engagement (Karasek, 1979). The ability to modify one's work independently reinforces the sense of agency and the belief that one has an impact on one's surroundings. The need for a positive self-image, as described among others in social identity theory (Tajfel, 1979), motivates actions that enhance one's self-worth and prestige, also through the way one defines one's work role. The third need is relationship building – strongly linked to a sense of belonging and meaningful work (Ryan & Deci, 2000). These relationships – for example, offering support to colleagues – are key to shared responsibility for safety within the team. Understanding these psychological underpinnings of job crafting allows one to see its broader context, including in the face of organisational change. Successful implementation of change requires consideration of the personal motivations and meanings given to work by employees. If organisational change corresponds to an individual's intrinsic needs – such as autonomy, meaning and relationality – then acceptance and active engagement are easier to achieve. In this view, job crafting can be seen as a way of adapting to change, enabling not only adaptation but also personal meaning. Self-modifying job crafting fosters a sense of influence and supports actions for security, both individual and community. Understanding these psychological underpinnings of job crafting allows us to see its wider context, including in the face of organisational change. As indicated in the previous chapter, effective implementation of change requires taking into account the personal motivations and meanings employees give to work. If organisational change corresponds to an individual's intrinsic needs – such as autonomy, meaning and relationality – then acceptance and active engagement are easier to achieve. In this view, job crafting can be seen as a way of adapting to change, enabling not only adaptation, but also personal meaning. Self-modifying the job fosters a sense of influence and supports actions for security, both individual and community. However, for this process to flourish, an organisational environment based on trust, openness and respect for autonomy is essential. A key condition is to move away from a blame culture that stifles initiative and leads to disengagement. Employees only take action beyond their formal responsibilities when they have the space to act without fear of being judged. It is the relational and cognitive dimension of job crafting, based on task redefinition and community building, that requires an atmosphere of psychological safety in which innovation and shared responsibility are encouraged rather than punished.

4.3 THE PRACTICAL ASPECTS OF IMPLEMENTING JOB CRAFTING IN THE CONTEXT OF SAFETY

Implementing the concept of job crafting in the context of job safety is an innovative yet practical approach to building a safety culture in an organisation. Unlike traditional top-down methods of risk control and management, this approach is based on strengthening the bottom-up potential of workers – their initiative, reflection and informed decisions about how they do their work. Job crafting assumes that each employee can – within a defined organisational framework – modify their tasks, their working relationships and the way they perceive their own role and its importance. In the context of safety, this means the real possibility of creating a safer working environment through personal involvement and the daily conscious shaping of behaviour.

Implementing job crafting, however, cannot be left solely to the individual efforts of employees. While its essence is voluntary and autonomous, an organisational environment – one that not only tolerates, but actively inspires, supports and recognises job crafting activities – plays a key role. This requires a shift in thinking among management: from the role of controller to that of facilitator and development partner. Management, through for example, communication and creating space for experimentation, shapes the climate in which job crafting can thrive. Practically, this can take place through, among other things, development programmes, workshops, individual and team coaching, reflection sessions or micro-interventions. It is also crucial to integrate job crafting into existing health and safety, HR or quality management initiatives so that it is not treated as a one-off project but becomes a permanent part of the organisational culture.

Job crafting can not only be a tool to support adaptation to change, but also an impetus for deep transformation and cultural reinforcement (cf. Figure 4.1). Employees' proactivity in shaping their own work – by modifying tasks, relationships and role perceptions – becomes a source of sustainable organisational change. In practice, this can mean improving the ergonomics of the job, developing cooperation on risky tasks or making work personally meaningful through its connection to the common good.

Activities of this kind trigger internal adaptation mechanisms. Employees who have a say in how they work experience a greater sense of control, satisfaction and psychological security. These psychological changes, in turn, foster the emergence of proactive behaviours, such as reporting risks, sharing ideas for improving working conditions, initiating changes in procedures or supporting colleagues. These behaviours have modelling potential and spread within the team, leading to an alignment of norms and patterns that support the development of a safety culture. In a mature

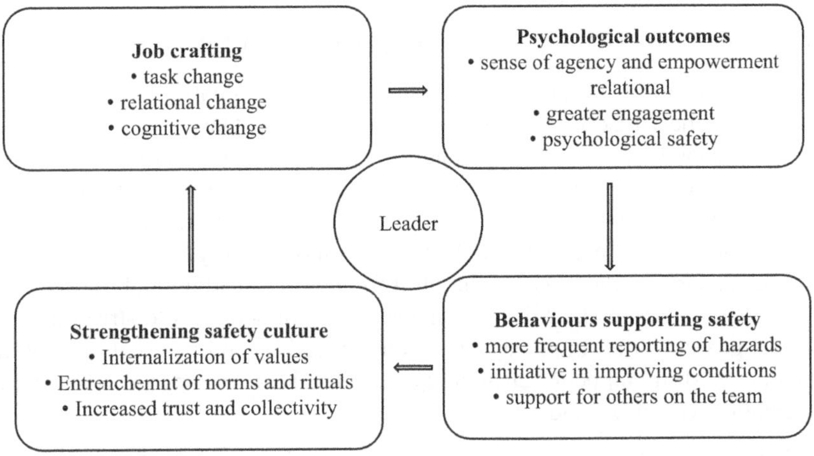

FIGURE 4.1 Model for strengthening safety culture through job crafting.

Source: Own elaboration.

organisation, safety culture is based not only on formal procedures, but also on realistically shared values and habits. Shared rituals are established, such as daily check-ins, rules for responding to risks, open communication about mistakes or mutual support. There is a growing level of trust, both horizontally, between employees, and vertically, in the relationship with leaders, which creates an environment conducive to further bottom-up initiatives and continuous improvement. At the centre of this process is the leader, a key figure in enhancing employees' proactivity and sense of influence. It is their attitude that determines whether initiatives are noticed, supported and integrated into organisational goals. A leader who promotes safety, models desirable behaviour and fosters autonomy has not only a controlling function, but – above all – also a developmental one.

Job crafting, thus, becomes not only an adaptive tool, but also a driver of transformation – enabling the seamless blending of individual actions with organisational values. In this view, safety ceases to be merely a set of rules and becomes a shared responsibility rooted in everyday decisions and relationships.

Of particular importance in the context of building a safety culture is working on three levels of job crafting: task crafting, relational crafting and cognitive crafting. Each of these levels can provide a viable starting point for shaping behaviours that foster safety – through both bottom-up action by employees and conscious support from leaders. The following are brief case studies that illustrate practical implementations of job crafting in the area of workplace safety. Each case concludes with a reflection question to encourage the reader to consider how similar actions could be taken in their work environment. Each of the cases described includes a reflective question that encourages personal reflection on how analogous solutions could be applied in the context of one's own job role or team.

CASE 1: I MODIFY A TASK TO MAKE IT SAFER

In a manufacturing plant, a leader asked the team to identify tasks that might cause fatigue, frustration or reduce concentration. During a joint session, employee Paul reported that cleaning a machine performed at the end of a shift involves a lot of tension – everyone is already tired but wants to finish the job as quickly as possible. Paul suggested moving this activity to earlier in the day – after the first break, when the team has more energy.

The manager agreed to pilot this shift for a week. In addition, he asked Paul to prepare a simple checklist with the sequence of activities to streamline the whole process. After three days, the team noticed less piling up of tasks at the end of the shift and better cleaning quality. The idea was approved for implementation as a new standard.

What small modifications to your daily work could improve safety?

CASE 2: I CREATE RELATIONSHIPS BASED ON SHARED CARE

In the packaging department, two employees – Emily and John – noticed that during the morning shifts, small errors often occurred due to rushing and unmentioned information. On their own initiative, they started talking for a few minutes each morning about what a challenge might be that day – for example, missed deliveries, equipment breakdowns, changes in documentation.

After a week, they noticed an improvement in the organisation of tasks and greater alertness to unusual situations. In a conversation with their manager, they suggested introducing similar "morning safety pairs" across the team. The manager incorporated the idea into pre-shift meetings, encouraging others to form two- to three-person micro-teams.

With whom in your environment could you create such a daily "safety micro-relationship"?

CASE 3: REDISCOVERING THE MEANING OF MY WORK

The leader of the technical team asked employees to list their three main tasks and describe how they affect the safety of others. Employee Jack, responsible for checking valves on pressure equipment, wrote: "This way, others can work safely without fear of failure".

During a meeting with the team, the leader quoted several anonymous statements, demonstrating the importance of each team member being attentive. Jack later admitted that after this exercise, he started to think about his role differently – he stopped treating it as "mere control" and started to see it as a real impact on the lives and health of colleagues.

Do you notice how your daily work contributes to the safety of others?

CASE 4: I CREATE SPACE FOR INITIATIVE – AS A LEADER

In the logistics department, manager Anna started a series of *Five Minutes for Innovation* – a weekly short meeting where everyone could put forward an idea to improve the tasks they perform. At the first meeting, she asked the question: "Which task do you perform regularly, and do you have ideas on how it could be done more safely or efficiently?"

Employees provided more than a dozen suggestions in writing. Two – concerning a change in the order of unloading and a different arrangement of boxes – were selected for piloting. After a week, a mini-survey was conducted, and the authors of the solutions reported what had worked. The ideas were incorporated into the team's practice.

As a leader, how can you support grassroots activities that affect safety and efficiency?

CASE 5: I INTRODUCE SAFETY MICRO-TEAMS

Due to an alarming increase in the number of minor incidents, shift leader Bob proposed the creation of "vigilance micro-teams" – groups of two to three people who were to share insights into potential risks and support each other on a daily basis.

Bob presented this not as an obligation, but as a form of mutual concern – "so that we don't overlook what might happen". The teams started meeting over coffee for three-minute talks. After a month, more than a dozen ideas were collected for improving the setting of the workspace and signage.

How can you strengthen relationships in your team based on trust and shared responsibility for safety?

CASE 6: HELPING PEOPLE SEE THE IMPORTANCE OF THEIR WORK

During individual development interviews, the department manager asked each employee two questions: Which of your tasks most affects the safety of others? What could happen if you did not perform it carefully?

The responses often included references to invisible but essential actions: locking machines, checking the status of tools and communicating when hand ing over a shift. Based on these conversations, the team collectively created a slogan: *Safety starts with what you can't see*.

The slogan was hung in the shared space, and its presence began to trigger regular conversations about what really "makes a difference" in day-to-day work.

How can you help yourself or others see the meaning and impact of everyday work?

The cases presented show that job crafting activities – at the level of tasks, relationships and cognitive interpretation of work – can be effectively developed in practice. What is important is that many of these initiatives have direct points of contact with the BBS approach, in which proactivity, attentiveness and bottom-up employee involvement in building a safe working environment are also central. In both the job crafting approach and the BBS model, worker proactivity plays a key role. In both cases, employees are not mere doers of tasks, but active participants in shaping their work environment. In BBS, the individual observes, reacts and takes action to improve safety – similar to job crafting, where the employee modifies the scope of responsibilities, the working relationship or the perception of the meaning of their work on their own. In this context, it can be considered that participating in the BBS process – especially after modifying the observation tools – can be a form of job crafting. An employee who observes behaviour, analyses it and initiates change actually transforms their work environment in response to their own needs, beliefs and observations. The BBS process involves the active participation of line employees as observers, whose involvement and collective decisions foster a sense of influence, agency and shared responsibility. It is these qualities – so important in job crafting – that develop through practising attentiveness to behaviour, conversations with others and daily reflection on safety.

CASE 7

In one workshop, a participant in the BBS programme admitted that after a few weeks of conducting behavioural observations, she began to notice risks not only at work, but also outside of work. "I started to pay attention to whether people were crossing the street in an unauthorised place. I became more attentive and vigilant myself – it stayed with me", she said. This shows how BBS activities can support the development of intrinsic motivation and self-awareness, the key mechanisms that drive job crafting.

In this way, BBS and job crafting are not only complementary, but can be mutually reinforcing – creating an environment where safety is not just an observed duty, but a conscious value co-created by all members of the organisation.

Juxtaposing these practical experiences with research findings, it is clear that developing a safety culture is important for employee well-being and organisational performance. Numerous studies show that a strong safety culture is associated with lower accident rates (Barling et al., 2002), better mental and physical health (Barling et al., 2002; Probst & Estrada, 2010) and higher levels of job satisfaction (Tamakloe et al., 2022). At the same time, there is a lack of in-depth research linking job crafting and safety culture – even though there is a strong theoretical and applied basis to develop this topic. Employees who are given the opportunity to shape their work show greater engagement (Tims et al., 2013), are more likely to be proactive (Ghitulescu, 2006) and better match tasks to their competencies, which reduces the risk of burnout and increases the willingness to share responsibility for safety (Petrou et al., 2012; Wrzesniewski & Dutton, 2001).

Implementing and supporting job crafting in organisations can, therefore, not only contribute to improving job quality, job satisfaction and employee engagement, but also provide a sustainable foundation for building a mature safety culture. This culture is not based solely on formal procedures but develops from the bottom up – through conscious attentiveness, a sense of meaning in the tasks performed and cooperation based on trust and mutual concern. By giving employees space for autonomy, initiative and co-creation of the working environment, the organisation not only reacts to risks, but actively prevents them. Thus, job crafting – especially when supported systemically by leaders and integrated with health and safety practices – becomes not only a tool for individual development, but also a mechanism for strengthening community responsibility for safety. This approach responds to the challenges of today's organisations, which are increasingly looking for agile, participatory methods of engaging employees. Job crafting in the context of safety is not just a personnel innovation; it is a practice that can realistically translate into increased organisational resilience and quality of working life.

4.4 CHALLENGES AND LIMITATIONS OF JOB CRAFTING IN A SAFETY CONTEXT

While job crafting offers a number of benefits, such as increased engagement, a sense of agency and well-being, its responsible implementation in the work environment is not without its challenges. This is particularly true in contexts where physical and psychological safety is one of the key dimensions of an organisation's daily functioning. In such contexts, not every bottom-up initiative – even if driven by positive intentions – can automatically be considered beneficial. Implementing job crafting in sectors heavily regulated by regulations, industry standards or quality requirements requires great care and often close integration with existing health and safety procedures. It becomes crucial here not only to create space for action, but also to clearly anchor these activities in organisational values and safety culture.

However, the challenges are not only about the formal framework. Simply allowing employees to influence the working environment does not guarantee that they will take advantage of this opportunity. The organisational climate – openness to dialogue, tolerance for experimentation and the level of psychological safety – is decisive. Only in an atmosphere of trust and a lack of fear of evaluation or punishment are employees willing to take initiatives, share observations and redefine their tasks.

In this context, it becomes crucial to move from a blame culture to a crafting culture, based on the recognition of the employee's role as a co-creator of the work environment rather than merely a recipient of systemic solutions. This approach requires a redefinition of power relations, a change in leadership style and the building of systems that reinforce autonomy and co-responsibility. Leaders have a special role here; as facilitators of change, they are the ones who model openness, recognise potential and support bottom-up initiatives. Leaders who do not act as catalysts for change, reinforcing initiative and autonomy, may unwittingly inhibit the development of crafting behaviour. Modelling desirable attitudes, celebrating small successes and valuing grassroots initiatives are also key. For many employees, a barrier may be a lack of understanding of the meaning of change; so, communicating its purpose and value is a matter not only of information, but also of building a cognitive basis for job crafting.

Another barrier can be insufficient psychological insecurity, which lowers the willingness to take risks, redefine roles or propose changes. Personalisation of the benefits proves to be equally important; if the change is not linked to the individual needs of the employees, it will be more difficult to gain their real commitment. In this context, job crafting can mitigate emotional resistance, transforming anxiety into activity and supporting psychological adaptation to changing conditions. Moreover, structural or technological changes can open up new opportunities for redefining roles and tasks, provided the organisation recognises and supports these bottom-up micro-transformations.

Ultimately, successfully combining top-down and bottom-up approaches – as the metaphor of the "transformation triangle" shows – requires not only systemic decisions, but also space for individual influence. This is where job crafting can become a force to support the implementation of change, as long as the organisation abandons a culture of blame in favour of a climate of trust, openness and shared responsibility.

It is also worth noting that job crafting and safety culture operate in the same organisational layers: they are rooted in values, norms and attitudes. Their effectiveness depends on the quality of relationships, the level of trust and how they are managed. Job crafting can be seen as a tool to support culture transformation – not only improving the fit between work and the individual but also reinforcing the daily focus on safety and well-being. Similarly, practices such as BBS show that engaging workers in the observation, reflection and improvement of working conditions leads to attentiveness that goes beyond formal procedures and becomes part of everyday action – bringing it closer to the essence of job crafting.

In order for job crafting to foster a safety culture in a responsible and sustainable way, it is worth considering the following key challenges and constraints:

1. **Formal and procedural constraints in high-risk environments**

In many industries – such as manufacturing, transport, energy or healthcare – performing tasks in accordance with health and safety procedures is not only a legal obligation, but also a prerequisite for life and health safety. Employees in such environments are not always free to "transform" tasks without prior consultation or testing. Job crafting, undertaken spontaneously and on a case-by-case basis, can generate risks in such contexts – especially if the changes involve how procedures are implemented, shortening work steps or reinterpreting risks. Organisations, therefore, need to set a clear *framework for safe crafting*, indicating which aspects of the work can be modified and which are inviolable.

2. **A lack of equal competence and sense of influence**

Job crafting is based on the assumption that the employee is willing and able to influence their work. However, the reality is that many people – especially those working in a strongly hierarchical culture or with low levels of autonomy – may not have a developed ability to reflect on their job role or may simply not believe that their actions matter. An underdeveloped sense of self-esteem, lack of confidence and fear of being judged – all of these can effectively block potential crafting activities. In such cases, the organisation must not only enable change, but also build psychological safety, that is, an environment where employees feel they can express ideas and concerns without fear of ridicule, rejection or negative consequences (Edmondson, 1999).

3. **A lack of trust in relationships and fear of evaluation**

Another constraint that can block job crafting is a lack of trust within the team and a fear of voicing opinions. Employees who do not feel psychologically safe rarely take initiatives beyond the formal minimum. In the context of safety, where signalling threats or the need to change a procedure can be perceived as criticism, psychological safety becomes the foundation for readiness to act. If an employee fears that their suggestions will be ignored, laughed at or used against them, they will not try to change anything – even if they perceive inefficiencies or risks.

4. **Conflict between individual initiative and team cohesion**

Job crafting is, by definition, an individual activity, whereas safety – both physical and psychological – is communal. This means that individual changes in the way of working can affect other team members, even if made in good faith. This can lead to tensions, misunderstandings or a lack of cohesion. For example, if one employee changes the order in which tasks are carried out, another – relying on this sequence – may lose focus. To prevent this, leaders need to promote *team dialogue around job crafting*, encouraging the sharing of ideas and agreeing on changes with the team.

5. **Invisibility of effects and difficulty in assessing impact**

Job crafting can be difficult to capture and measure – its outcomes are often qualitative, subjective and dispersed over time. In a safety context, where hard metrics (number of accidents, reports or inspections) dominate, effects related to improved well-being, increased attentiveness or improved relationships are sometimes ignored or underestimated. Meanwhile, it is these soft factors that influence the willingness to follow procedures, report risks or support others. Organisations should, therefore, create *hybrid performance assessment systems* that combine quantitative data (e.g. incident rates) with qualitative data (e.g. levels of sense of impact, collaboration, sense of work).

6. **The need for continuity and structural support**

Job crafting cannot be treated as a one-off campaign or part of a training project. It is a process that requires a *long-term approach*, consistent support from leaders and being embedded in a broader organisational strategy. Employees need to be assured that their initiatives will not only be listened to but also incorporated into the company's safety policy wherever possible. Failure to respond to their involvement can lead to a decline in motivation and a sense of lack of influence – the opposite of what is intended.

While job crafting opens up huge opportunities for organisations to build commitment and shared responsibility, its implementation – especially in the context of safety – requires caution, awareness and a readiness to support employees at multiple levels. The key here is a combination of:

- clear procedural boundaries,
- a culture of openness and dialogue,
- a high level of *psychological safety* that allows people to talk openly about needs, ideas and concerns.

Job crafting is not a panacea – it is a tool that only works in organisations that are ready for trust, flexibility and shared learning. Its potential to build a culture of safety is enormous if accompanied by systemic support and informed leadership.

References

Adjekum, D. K., & Terns, M. F. (2020). Assessing the relationship between organizational management factors and a resilient safety culture in a collegiate aviation program with Safety Management Systems (SMS). *Safety Science*, *13*. https://doi.org/10.1016/j.ssci.2020.104909

Alavosius, M., & Burleigh, K. (2021). Behavior-based safety as a replicable technology. *Applied Behavior Science in Organizations*. https://doi.org/10.4324/9781003198949-2

Al-Bayati, A. J., Abreu, T. T. P. D., & Ko, C. H. (2025). Tackling challenges in implementing behavior-based safety (BBS): Construction safety culture focus. *Journal of Engineering, Project, and Production Management*, *15*(1). https://doi.org/10.32738/JEPPM-2025-0005

Al-Haddad, S., & Kotnour, T. (2015). Integrating the organizational change literature: A model for successful change. *Journal of Organizational Change Management*, *28*(2), 234–262. https://doi.org/10.1108/JOCM-11-2013-0215

Ansoff, H. I. (1985). *Zarządzanie strategiczne* [Strategic Management]. PWE.

Appelbaum, S., Cameron, A., Ensink, F., Humphries, E., MacLeod, E., & Meicke, M. (2017). Factors that impact the success of an organizational change: A case study analysis. *Industrial and Commercial Training*, *49*(4), 213–230. https://doi.org/10.1108/ICT-02-2017-0006

Auzoult, L., & Gangloff, B. (2018). The mediating role of integration of safety by activity versus operator between organizational culture and safety climate. *International Journal of Injury Control & Safety Promotion*, *25*(4), 433–438. https://doi.org/10.1080/17457300.2018.1456471

Ayuni, M. Q., Yusuf, M., & Dwiyanti, E. (2022). Performance analysis of the behavior based safety program in reducing occupational accident rates. *The Indonesian Journal of Occupational Safety and Health*, *11*(2), 275–284. https://doi.org/10.20473/ijosh.v11i2.2022.275-284

Bae, H., Simmons, D. R., & Polmear, M. (2021). Promoting the quarry Workers' Hazard identification through formal and informal safety training. *Safety and Health at Work*, *12*(3), 317–323. https://doi.org/10.1016/j.shaw.2021.02.003

Bakker, A. B., & Bal, P. M. (2010). Weekly work engagement and performance: A study among starting teachers. *Journal of Occupational and Organizational Psychology*, *83*, 189–206. http://dx.doi.org/10.1348/096317909X402596

Bakker, A. B., Demerouti, E., & Verbeke, W. (2004). Using the job demands-resources model to predict burnout and performance. *Human Resource Management*, *43*(1), 83–104. https://doi.org/10.1002/hrm.20004

Baran, M., & Sypniewska, B. (2020). The impact of management methods on employee engagement. *Sustainability*, *12*(1). https://doi.org/10.3390/su12010426

Barling, J., Loughlin, C., & Kelloway, E. K. (2002). Development and test of a model linking safety-specific transformational leadership and occupational safety. *Journal of Applied Psychology*, *87*(3), 488–496. https://doi.org/10.1037/0021-9010.87.3.488

Bayramova, A., Edwards, D. J., Roberts, C., & Rillie, I. (2023). Constructs of leading indicators: A synthesis of safety literature. *Journal of Safety Research*, *85*, 469–484. https://doi.org/10.1016/j.jsr.2023.04.015

Berg, J. M., Dutton, J., & Wrzesniewski, A. (2021). *What is job crafting and why does it matter? Michigan Ross School of Business 2008. Center for Positive Organizations* [Internet]. https://positiveorgs.bus.umich.edu/wp-content/uploads/

Berg, J. M., Dutton, J. E., & Wrzesniewski, A. (2013). Job crafting and meaningful work. In B. J. Dik, Z. S. Byrne, & M. F. Steger (Eds.), *Purpose and meaning in the workplace*, 81–104. American Psychological Association. https://doi.org/10.1037/14183-005

Bergheim, K., Nielsen, M. B., Mearns, K., & Eid, J. (2015). The relationship between psychological capital, job satisfaction, and safety perceptions in the maritime industry. *Safety Science*, *74*, 27–36. https://doi.org/10.1016/j.ssci.2014.11.024

Boczkowska, K., Niziołek, K., & Roszko-Wójtowicz, E. (2022). A multivariate approach towards the measurement of active employee participation in the area of occupational health and safety in different sectors of the economy. Equilibrium. *Quarterly Journal of Economics and Economic Policy*, *17*(4), 1051–1085. https://doi.org/10.24136/eq.2022.035

Boczkowska, K., & Znajmiecka-Sikora, M. (2015). Behavior-based safety – as the effective anti-crisis tool. *Marketing i Rynek*, *5*. http://bazekon.icm.edu.pl/bazekon/element/bwmeta1.element.ekon-element-000171360351

Burnes, B. (2011). Introduction: Why does change fail, and what can we do about it? *Journal of Change Management*, *11*(4), 445–450. https://doi.org/10.1080/14697017.2011.630507

Bye, R. J., Aalberg, A. L., & Royrvik, J. O. D. (2020). What we talk about when we talk about HSE and culture – a mapping and analysis of the academic discourses. *Safety Science*, *129*. https://doi.org/10.1016/j.ssci.2020.104846

Carra, S., Bottani E., Vignali, G., Madonna, M., & Monica L. (2023). Implementation of behavior-based safety in the workplace: A Review of conceptual and empirical literature. *Sustainability*, *16*. https://doi.org/10.3390/su162310195

Casey, T. W., Hu, X., Reid, C., Tran, P. A., & Guldenmund, F. W. (2022). Rolling up our sleeves and pulling up our socks: A critical review of safety culture definitions and measures, and innovative ways to move the field forward. In *Handbook of Research Methods for Organisational Culture; School of Management and Ruth Knight, Postdoctoral Research Fellow, Australian Centre for Philanthropy and Nonprofit Studies*, Queensland University of Technology, Brisbane, Australia. ISBN 9781788976251. https://www.e-elgar.com/shop/gbp/handbook-of-research-methods-for-organisational-culture-9781788976251.html

Chen, D., & Tian, H. (2012). Behavior based safety for accidents prevention and positive study in China construction project. *Procedia Engineering*, *43*, 528–534. https://doi.org/10.1016/j.proeng.2012.08.092

Chen, Y., McCabe, B., & Hyatt, D. (2017). Impact of individual resilience and safety climate on safety performance and psychological stress of construction workers: A case study of the Ontario construction industry. *Journal of Safety Research*, *61*, 167–176. https://doi.org/10.1016/j.jsr.2017.02.014

Choudhry, R. M., Fang, D., & Mohamed, S. (2007). The nature of safety climate: A survey of the state-of-the-art. *Safety Science*, *45*, 993–1012. https://doi.org/10.1016/j.ssci.2006.09.003

Choudhry, R. M. (2014). Behavior-based safety on construction sites: A case study. *Accident Analysis & Prevention*, *70*, 14–23. https://doi.org/10.1016/j.aap.2014.03.007

Clarke, L. (1997). *Zarządzanie zmianą [Managing change]*. Gebethner & Ska.

Colley, A. (2013). An invisible population? The needs of young women offenders and why gender deserves consideration in the Aotearoa New Zealand Youth Justice system. *NZLSJ*, *3*, 471–487. https://nzlsj.com/wp-content/uploads/2016/07/3-allanah-colley-2015-3-nzlsj-471.pdf

Cooper, D. (1998). *Improving safety culture: A practical guide*. Wiley. https://www.behavioral-safety.com/articles/Improving_safety.culture_a_practical_guide.pdf

Cooper, D. (2009). *Behavioral safety: A framework for success*. Behavioral Safety Services.

Cooper, M. D. (2000). Towards a model of safety culture. *Safety Science*, *36*(2), 111–136. https://www.nrc.gov/docs/ML0906/ML090640285.pdf

Daniels, A., & Bailey, J. (2014). *Performance management: Changing behavior that drives organizational effectiveness*. Aubrey Daniels International. https://dokumen.pub/qdownload/performance-management-changing-behavior-that-drives-organizational-effectiveness-5nbsped-9780937100257-b-1352039.html

Dedobbeleer, N., & Béland, F. (1991). A safety climate measure for construction sites. *Journal of Safety Research*, *22*(2), 97–103. https://doi.org/10.1016/0022-4375(91)90017-P

Dobrowolska, M., Stasiła-Sieradzka, M., & Kozuba, J. (2020). Safety attitude as a predictor of the sense of threat in the workplace, using the example of airport ground staff. *Sustainability*, *12*. https://doi.org/10.3390/su12166569

Donald, I., & Canter, D. (1994). Employee attitudes and safety in the chemical industry. *Journal of Loss Prevention in the Process Industries*, *7*(3), 203–208. https://doi.org/10.1016/0950-4230(94)80067-7

Dubbelt, L., Demerouti, E., & Rispens, S. (2019). The value of job crafting for work engagement, task performance, and career satisfaction: Longitudinal and quasi-experimental evidence. *European Journal of Work and Organisational Psychology*, *28*(3). https://doi.org/10.1080/1359432X.2019.1576632

Edmondson, A. C. (1999). Psychological safety and learning behavior in work teams. *Administrative Science Quarterly*, *44*(2), 350–383. https://doi.org/10.2307/2666999

Edwards, J., Freeman, J., Soole, D., & Watson, B. (2014). A framework for conceptualising traffic safety culture. *Transportation Research Part F: Traffic Psychology and Behaviour*, *26*(1), 293–302. https://doi.org/10.1016/j.trf.2014.03.002

Errida, A., & Lotfi, B. (2021). The determinants of organizational change management success: Literature review and case study. *International Journal of Engineering Business Management*, *13*. https://doi.org/10.1177/18479790211016273

Esitikot, E., Umoh, M., Ekong, A., Okeke, G., Ofon, U., Akadi, A., & Obadimu, C. (2025). Behaviour-based safety as a strategic instrument for accident reduction in the Nigerian petroleum industry. *Asian Research Journal of Arts & Social Sciences*, *23*(4). https://doi.org/10.9734/arjass/2025/v23i4666

Fatolitis, P. G. (2018). Behavior-based safety program effectiveness and culture: A brief review and update. *Journal of Ergonomics*, *8*(E182). https://doi.org/10.4172/2165-7556.1000E182

Fazilov, S., Senouci, A., & Al-Derham, H. (2018). Using behavioral based safety to enhance safety performance. *Journal of Civil Engineering and Construction*, *7*(4), 192–201. https://doi.org/10.32732/jcec.2018.7.4.192

Fleming, M. (2005). Patient safety culture measurement and improvement: A 'how to' guide. *Healthcare Quarterly*, *8*(Spec No), 14–19. https://doi.org/10.12927/hcq.2005.17656

Flin, R., McAms, K., O'Connor, P., & Bryden, R. (2000). Measuring safety climate: Identifying the common features. *Safety Science*, *34*(1–3), 177–192. https://doi.org/10.1016/S0925-7535(00)00012-6

Frederick, D. E., & VanderWeele, T. J. (2020). Longitudinal meta-analysis of job crafting shows positive association with work engagement. *Cogent Psychology*, *7*(1). https://doi.org/10.1080/23311908.2020.1746733

Fu, Y. K., & Chan, T. L. (2014). A conceptual evaluation framework for organisational safety culture: An empirical study of Taipei Songshan Airport. *Journal of Air Transport Management*, *34*, 101–108. https://doi.org/10.1016/j.jairtraman.2013.08.005

Gabryelewicz, I., & Krupa, P. (2016). Poziom klimatu bezpieczeństwa jako element zarządzania bezpieczeństwem w przedsiębiorstwie [The level of safety climate as an element of safety management in an enterprise]. *Zeszyty Naukowe Politechniki Częstochowskiej. Zarządzanie*, *19*, 183–192. https://www.sbc.org.pl/Content/358714/gabrylewicz.pdf

Galloway, S. M. (2011). The contributing factors of behavior-based safety failures. *EHS Today*. https://proactsafety.com/articles/contributing-factors-of-bbs-failures

Geller, E. S. (2000). *The psychology of safety handbook*. (2nd ed.). CRC Press.

Geller, E. S. (2005). Behavior-based safety and occupational risk management. *Behavior Modification*, *29*(3), 539–561. https://doi.org/10.1177/0145445504273287

Ghitulescu, B. (2006). Shaping tasks and relationships at work: Examining the antecedents and consequences of employee job crafting. http://dscholarship.pitt.edu/10312/1/ghitulescube_etd.pdf

Gillen, M., Baltz, D., Gassel, M., Kirsch, L., & Vaccaro, D. (2002). Perceived safety climate, job demands, and coworker support among union and nonunion injured construction workers. *Journal of Safety Research*, *33*(1), 33–51. https://doi.org/10.1016/S0022-4375(02)00002-6

Golubovich, J., Chang, C.-H., & Eatough, E. M. (2014). Safety climate, hardiness, and musculoskeletal complaints: A mediated moderation model. *Applied Ergonomics*, *45*(3), 757–766. https://doi.org/10.1016/j.apergo.2013.10.008

Goncalves Filho, A. P., & Wotcrson, P. (2018). Maturity models and safety culture: A critical review. *Safety Science*, *105*, 192–211. https://doi.org/10.1016/j.ssci.2018.02.017

Griffin, R. W. (2005). *Podstawy zarządzania organizacjami* [Fundamentals of organizational management]. Wydawnictwo Naukowe PWN.

Håvold, J. I. (2010). Safety culture and safety management aboard tankers. *Reliability Engineering & System Safety*, *95*, 511–519. https://doi.org/10.1016/j.ress.2010.01.002

Hofstede, G. (2000). *Kultury i organizacje: Zaprogramowanie umysłu* [Cultures and organizations: Programming the mind]. PWN.

Hopkins, A. (2019). *Organising for safety*. Wolters Kluwer.

Hu, Q., Wang, H., & Long, L. (2020). Will newcomer job crafting bring positive outcomes? The role of leader-member exchange and traditionality. *Acta Psychologica Sinica*, *52*(5). https://doi.org/10.3724/SP.J.1041.2020.00659

Hudson, P. (2007). Implementing a safety culture in a major multinational. *Safety Science*, *45*(6), 697–722. https://doi.org/10.1016/j.ssci.2007.04.005

Hunter, S. T., Bedell, K. E., & Mumford, M. D. (2007), Climate for creativity: A quantitative review. *Creativity Research Journal*, *19*(1), 69–90. https://doi.org/10.1080/10400410709336883

IAEA (1991). *Safety culture*. International Safety Advisory Group, Safety Series 75INSAG4.

International Organization for Standardization. (2018). ISO 45001:2018 *Occupational health and safety management systems — Requirements with guidance for use. International Organization for Standardization*. https://www.iso.org/standard/63787.html

Jacobs, G., van Witteloostuijn, A., & Christe-Zeyse, J. (2013). A theoretical framework of organizational change. *Journal of Organizational Change Management*, *26*(5), 772–792. https://doi.org/10.1108/JOCM-09-2012-0137

Kandathil, G., & Joseph, J. (2019). Normative underpinnings of direct employee participation studies and implications for developing ethical reflexivity: A multidisciplinary review. *Journal of Business Ethics*, *157*, 685–697. https://doi.org/10.1007/s10551-017-3689-x

Kanter, R. M. (2011). *The change wheel: Elements of systemic change and how to get change rolling* (Background Note No. 312-083). Harvard Business School Press.

Kapica, Ł., & Baka, Ł. (2021). *Czym jest job crafting? Przegląd koncepcji teoretycznych dotyczących kształtowania pracy* [What is job crafting? A review of theoretical concepts related to job design]. *Medycyna Pracy*, *72*(4), 423–436. https://doi.org/10.13075/mp.5893.01115

Kaplan, R. S. (1996). *The balanced scorecard: Translating strategy into action*. Harvard Business School Press.

Karasek, R. A. (1979). Job demands, job decision latitude, and mental strain: Implications for job redesign. *Administrative Science Quarterly*, *24*(2). http://dx.doi.org/10.2307/2392498

Kasprzak, E., Michalak, M., & Minda, M. (2017). *Kwestionariusz kształtowania pracy – KKPracy. Polska adaptacja narzędzia* [Job Crafting Questionnaire – KKPracy. Polish adaptation of the tool]. *Psychologia Społeczna*, *12*(4), 459–475.

Katz, D., & Kahn, R. (1979). *Społeczna psychologia organizacji* [Social Psychology of Organization]. PWN.

Kennedy, R., & Kirwan, B. (1998). Development of a Hazard and operability based method for identifying safety management vulnerabilities in high risk systems. *Safety Science*, *30*, 249–274. https://doi.org/10.1016/S0925-7535(98)00025-3

Komaki, J., Heinzmann, A. T., & Lawson, L. (1980). Effects of training and feedback: Component analysis of a behavioral safety program. *Journal of Applied Psychology*, *65*(3), 261–270. https://doi.org/10.1037/0021-9010.65.3.261

Kotter, J. P. (1995). Leading change: Why transformation efforts fail. *Harvard Business Review*, *73*(2), 59–67.

Koźlik, M. (2008). Obserwator – kluczowa rola w procesie BBS [The observer – key role in the BBS process]. *Promotor*, *30*, 30–31.

Koźmiński, A. K., & Piotrowski, W. (2006). *Zarządzanie. Teoria i praktyka* [Management: Theory and practice]. PWN.

Kożuch, B. (2011). *Nauka w organizacji [Knowledge in organization].* Wydawnictwo Fachowe CeDeWu.

Kożusznik, B. (2002). *Zachowania człowieka w organizacji* [Human behavior in organizations]. Polskie Wydawnictwo Ekonomiczne.

Krause, T. (1996). *The behavior-based safety process: Managing involvement for an injury-free culture*. Van Nostrand Reinhold.

Le Coze, J. C. (2019). How safety culture can make us think. *Safety Science*, *118*, 221–229. https://doi.org/10.1016/j.ssci.2019.05.026

Leana, C., Appelbaum, E., & Shevchuk, I. (2009). Work process and quality of care in early childhood education: The role of job crafting. *Academy of Management Journal*, *52*(6). https://doi.org/10.5465/amj.2009.47084651

Leavitt, H. J. (1964). Applied organization change in industry: Structural, technological and humanistic approaches. In W. J. G. March (Red.), *Handbook of organizations* (Chapter 5, s. 2118-2145). Routledge.

Lee, T., & Harrison, K. (2000). Assessing safety culture in nuclear power stations. *Safety Science*, *34*(1–3), 61–97. https://doi.org/10.1016/S0925-7535(00)00007-2

Likert, R. (1967). *The human organization: Its management and values*. McGraw-Hill.

Lundell, M. A. & Marcham, C. L. (2018). Leadership's effect on safety culture. *Professional Safety*, *63*(11). https://aeasseincludes.assp.org/professionalsafety/pastissues/063/11/F2LundellMarcham_1118.pdf

Mao, J. Y., Xiao, J., Liu, X., Qing, T., & Xu, H. (2022). Emulating coworkers: How and when coworker ideation facilitates employee ideation. *Creativity Research Journal*, *35*(1), 99–115. https://doi.org/10.1080/10400419.2022.2049533

Mark, N., Reader, T., Shorrock, S., & Kirvan, B. (2015). The relationship between national culture and safety culture: Implications for international safety culture assessments. *Journal of Occupational and Organizational Psychology*, *89*(3), 515–538. https://doi.org/10.1111/joop.12139

Markowski, A. (2022). *Bezpieczeństwo procesów przemysłowych* [Industrial process safety]. Wydawnictwo Politechniki Łódzkiej.

McSween, T. E. (2003). *The values-based safety process* (2nd ed.). John Wiley & Sons. https://content.e-bookshelf.de/media/reading/L-583717-2074327f3b.pdf

Mearns, K., Whitaker, S. M., & Flin, R. (2003). Safety climate, safety management practice and safety performance in offshore environments. *Safety Science*, *41*(8), 641–680. https://doi.org/10.1016/S0925-7535(02)00011-5

Mento, A. J., Jones, R. M., & Dirndorfer, W. (2002). A change management process: Grounded in both theory and practice. *Journal of Organizational Change Management*, *15*(1), 45–59. https://doi.org/10.1080/714042520

Mikkelsen, A., & Saksvik, P. (2004). The relationship between systematic OHS management and sick leave. *Journal of Occupational Health and Safety–Australia/New Zealand*, *20*(2), 169–179.

Milczarek, M. (2002). *Kultura bezpieczeństwa pracy* [Occupational safety culture] (Doctoral dissertation). CIOP – Centralny Instytut Ochrony Pracy.

Milczarek, M., & Najmiec, A. (2004). The relationship between workers' safety culture and accidents, near accidents and health problems. *International Journal of Occupational Safety and Ergonomics*, *10*(1), 25–33. https://doi.org/10.1080/10803548.2004.11076592

Minda, M., & Kasprzak, E. (2018). Czynniki kontekstu pracy i kształtowanie pracy jako predyktory przywiązania do organizacji [Work context factors and job crafting as predictors of organizational commitment]. *Organizacja i Kierowanie*, *3*, 145–161.

Nawakitphaitoon, K., & Zhang W. (2021). The effect of direct and representative employee voice on job satisfaction in China: Evidence from employer–employee matched data. *The International Journal of Human Resource Management*, *32*(22), 4725–4751. https://doi.org/10.1080/09585192.2020.1744028

Neal, A., & Griffin, M. A. (2006). A study of the lagged relationships among safety climate, safety motivation, safety behavior, and accidents at the individual and group levels. *Journal of Applied Psychology*, *91*(4), 946–953. https://doi.org/10.1037/0021-9010.91.4.946

Noort, M. C., Reader, T. W., Shorrock, S., & Kirwan, B. (2016). The relationship between national culture and safety culture: Implications for international safety culture assessments. *Journal of Occupational and Organizational Psychology*, *89*(3), 515–538. https://doi.org/10.1111/joop.12139

Nowacki, K., Łakomy, K., & Lis, T. (2016). Staż pracy a kultura bezpieczeństwa [Job tenure and safety culture]. In R. Knosala (Ed.), *Innowacje w zarządzaniu i inżynierii produkcji* [Innovations in management and production engineering] (pp. 448–460). Oficyna Wydawnicza Polskiego Towarzystwa Zarządzania Produkcją.

Nunu, W. N., Kativhu, T., & Moyo, P. (2019). An evaluation of the effectiveness of the Behaviour Based Safety Initiative card system at a cement manufacturing company in Zimbabwe. *Safety and Health at Work*, *10*(3), 231–238. https://doi.org/10.1016/j.shaw.2017.09.002

Olive, C., O'Connor, T. M., & Mannan, M. S. (2006). Relationship of safety culture and process safety. *Journal of Hazardous Materials*, *130*(1–2), 133–140. https://doi.org/10.1016/j.jhazmat.2005.07.043

Oprea, B. T., Barzin, L., Virga, D., Iliescu, D., & Rusu, A. (2019). Effectiveness of job crafting interventions: A meta-analysis and utility analysis. *European Journal of Work and Organisational Psychology*, *28*(6). https://doi.org/10.1080/1359432X.2019.1646728

Penc, J. (2000). Sprawne kierowanie zmianami. [Efficient change management] Annales Universitatis Mariae Curie-Skłodowska. *Sectio H, Oeconomia*, *34*, 185–195. https://bazhum.muzhp.pl/media/texts/annales-universitatis-mariae-curie-skodowska-sectio-h-oeconomia/2000-tom-34/annales_universitatis_mariae_curie_sklodowska_sectio_h_oeconomia-r2000-t34-s185-195.pdf

Petitta, L., Probst, T. M., Barbaranelli, C., & Ghezzi, V. (2017). Disentangling the roles of safety climate and safety culture: Multi-level effects on supervisor enforcement and employee safety compliance. *Accident Analysis & Prevention*, *99*, 77–89. https://doi.org/10.1016/j.aap.2016.11.012

Petrou, P., Demerouti, E., Peeters, M. C., Schaufeli, W. B., & Hetland, J. (2012) Crafting a job on a daily basis: Contextual correlates and the link to work engagement. *Journal of Organizational Behavior*, *33*. https://doi.org/10.1002/job.1783

Pidgeon, N. F. (1991). Safety culture and risk management in organizations. *Journal of Cross–Cultural Psychology*, *22*(1), 129–140. https://doi.org/10.1177/0022022191221009

Probst, T. M., Estrada, A. X. (2010). Accident under-reporting among employees: Testing the moderating influence of psychological safety climate and supervisor enforcement of safety practices. *Accident Analysis & Prevention*, *42*(5). https://doi.org/10.1016/j.aap.2009.06.027

Rajput, K., & Makwana, D. (2025). Evaluating the effectiveness of behavioural-based safety measures amongst workers working in industrial workplaces. *International Journal of Research Publication and Reviews*, *6*(4), 1339–1344. https://doi.org/10.55248/gengpi.6.0425.1349

Rashid, M. I., Athar, M., Noor, F., & Hussain, A. (2022). Behavior-based safety program for process industries. *International Journal of Occupational Safety and Ergonomics*, *29*. https://doi.org/10.1080/10803548.2022.2135282

Ratajczak, Z. (1980). *Człowiek w sytuacji innowacyjnej* [The human in an innovative situation]. PWN.

Roczniewska, M., Rogala A., Puchalska-Kaminska, M., Cieślak, R., & Retowski, S. (2020). I believe I can craft! introducing Job Crafting Self-Efficacy Scale (JCSES). *PLoS ONE*, *15*(8). https://doi.org/10.1371/journal.pone.0237250

Rogala, A., & Cieślak, R. (2019). Narzędzie do pomiaru przekształcania pracy: Właściwości psychometryczne polskiej wersji Job Crafting Scale [A tool for measuring job crafting: Psychometric properties of the Polish version of the Job Crafting Scale]. *Medycyna Pracy*, 70(4), 445–457. https://doi.org/10.13075/mp.5893.00822.

Rudolph, C. W., Katz, I. M., Lavigne, K. N., & Zacher, H. (2017) Job crafting: A meta-analysis of relationships with individual differences, job characteristics, and work outcomes. *Journal of Vocational Behavior*, *102*. https://doi.org/10.1016/j.jvb.2017.05.008

Ryan, R. M., & Deci, E. L. (2000). Self-determination theory and the facilitation of intrinsic motivation, social development, and well-being. *American Psychologist*, *55*(1), 68–78. https://doi.org/10.1037/0003-066X.55.1.68

Schein, E. H. (2010). *Organizational culture and leadership* (2nd ed.). John Wiley & Sons.

Schneider, B., GonzálezRomá, V., Ostroff, C., & West, M. A. (2017). Organizational climate and culture: Reflections on the history of the constructs in the Journal of Applied Psychology. *Journal of Applied Psychology*, *102*(3), 468–482. https://doi.org/10.1037/apl0000090.supp

Schwatka, N. V., Hecker, S., Lindsell, C. J., & Goldenhar, L. M. (2016). Defining and measuring safety climate: A review of the construction industry literature. *Annals of Occupational Hygiene*, *60*(5), 537–550. https://doi.org/10.1093/annhyg/mew020

Siu, O., Phillips, D. R., & Leung, T. (2004). Safety climate and safety performance among construction workers in Hong Kong: The role of psychological strains as mediators. *Accident Analysis and Prevention*, *36*(3), 359–366. https://doi.org/10.1016/S0001-4575(03)00016-2

Skinner, B. F. (1953). *Science and human behavior*. Macmillan. https://www.bfskinner.org/newtestsite/wp-content/uploads/2014/02/ScienceHumanBehavior.pdf

Slemp, G. R., Vella-Brodrick, D. A. (2013). The Job Crafting Questionnaire: A new scale to measure the extent to which employees engage in job crafting. *International Journal of Wellbeing*, *3*(2).

Smith, R., King, D., Sidhu, R., & Skryme, R. (2014). *The effective change manager's handbook: Essential guidance to the change management body of knowledge* (1st ed.). Kogan Page Ltd.

Sokołowska, S. (2009). *Organizacja i zarządzanie. Ujęcie teoretyczne* [Organization and management: A theoretical approach]. Wydawnictwo Uniwersytetu Opolskiego.

Stanisławski,R.(2017).*Triangulacjatechnikbadawczychwnaukachozarządzaniu*[Triangulation of research techniques in management sciences]. *Organizacja i Kierowanie*, *4*(178), 103–120.

Stasiła-Sieradzka, M., Chudzicka-Czupała, A., & Znajmiecka-Sikora, M. (2020) Work safety climate. Comparison of selected occupational groups. *PLoS ONE*, *15*(12). https://doi.org/10.1371/journal.pone.0243056

Stasiła-Sieradzka, M., & Dobrowolska, M. (2016). Badanie klimatu bezpieczeństwa pracy jako element benchmarkingu w zarządzaniu bezpieczeństwem w organizacji [Study of work safety climate as a benchmarking element in safety management in an organization]. *Społeczeństwo i Edukacja. Międzynarodowe Studia Humanistyczne*, *2*(21), 471–487.

Stasiła-Sieradzka, M., Znajmiecka-Sikora, M. (2017). Staff assessment of the work safety climate in its importance to human resource management in creating a safe work environment: The cooperative range of activities in human resource management end occupational health and safety. *Zarządzanie Zasobami Ludzkimi*, *5*(118):39–53.

Studenski, R. (1999). *Psychologia pracy* [Work psychology]. Politechnika Warszawska, Warszawa.

Studenski, R. (2000). Kultura bezpieczeństwa pracy [Occupational Safety Culture] w przedsiębiorstwie. *Bezpieczeństwo Pracy*, *9*, 1–4.

Szczupaczyński, J. (1995). *Podstawy zarządzania i kierowania ludźmi w organizacji* [Fundamentals of management and leadership in organizations]. Międzynarodowa Szkoła Menedżerów.

Szczupaczyński, J. (1998). *Anatomia zarządzania organizacją*. [Anatomy of organizational Management]. Międzynarodowa Szkoła Menedżerów.

Tajfel, H. (1979). Individuals and groups in social psychology. *British Journal of Social & Clinical Psychology*, *18*(2), 183–190. https://doi.org/10.1111/j.2044-8260.1979.tb00324.x

Tamakloe, S. A., Amponsah-Tawiah, K., & Mensah, J. (2022). Safety climate and psychological well-being among workers in the Ghanaian aviation industry: Does fatigue matter? *Occupational Health Science*, *6*(2), 279–293. https://doi.org/10.1007/s41542-022-00110-2

Thakare, R., & Singh, O. S. (2022). Study of obstacles in implementation of behaviour based safety system in an industry. *International Journal for Research Trends and Innovation*, *7*(10). https://www.ijrti.org/papers/IJRTI2210054

Thakre, N., & Shroff, N. (2016). Organizational climate, organizational role stress and job satisfaction among employees. *Journal of Psychosocial Research*, *11*(2), 469–478.

Tims, M., Bakker, A. B., & Derks, D. (2012). Development and validation of the job crafting scale. *Journal of Vocational Behavior*, 80(1): 173–186, https://doi.org/10.1016/j.jvb.2011.05.009

Tims, M., Bakker, A. B., & Derks, D. (2013). The impact of job crafting on job demands, job resources, and well-being. *Journal of Occupational Health Psychology*, *18*, 230–240. https://doi.org/10.1037/a0032141

Tims, M., Bakker, A. B., & Derks, D. (2015). Job crafting and job performance: A longitudinal study. *European Journal of Work and Organizational Psychology*, *24*(6), 914–928. https://doi.org/10.1016/j.jvb.2011.05.009

Tuncel, S., Lotlikar, H., Salem, O., & Daraiseh, N. (2006). Effectiveness of behaviour based safety interventions to reduce accidents and injuries in workplaces: Critical appraisal and meta-analysis. *Theoretical Issues in Ergonomics Science*, *7*(3), 191–209. https://doi.org/10.1080/14639220500090273

Tyszkiewicz, R. (2019). Istota kultury bezpieczeństwa pracy w systemie zarządzania [The essence of occupational safety culture in the management system]. *Doskonalenie jakości produkcji*, *2*, 94–101. https://www.qpi-journal.pl/EN/files/QPI_11_2019_08.pdf

Van Nunen, K., Li, J., Reniers, G., & Ponnet, K. (2018). Bibliometric analysis of safety culture research. *Safety Science*, *108*, 248–258. https://doi.org/10.1016/j.ssci.2017.08.011

Varonen, U., & Mattila, M. (2000). The safety climate and its relationship to safety practices, safety of the work environment and occupational accidents in eight wood-processing companies. *Accident Analysis and Prevention*, *32*(6), 761–769. https://doi.org/10.1016/S0001-4575(99)00129-3

Wang, B., Wu, C., & Huang, L. (2018). Emotional safety culture: A new and key element of safety culture. *Process Safety Progress*, *37*(2), 134–139.

Weaver, B., Kirk-Brown, A., Goodwin, D., & Oxley, J. (2023). Psychosocial safety behavior: A scoping review of behavior-based approaches to workplace psychosocial safety. *Journal of Safety Research*, *85*, 33–40. https://doi.org/10.1016/j.jsr.2022.10.006

Wendt, R. (2010). *Zarządzanie zmianą w polskiej firmie* [Change management in a Polish company]. Dom Wydawniczy Zachorek.

World Health Organization. (1946). *Preamble to the constitution of the world health organization as adopted by the international health conference*, New York, 19–22 June 1946; signed on 22 July 1946 by the representatives of 61 States (Official Records of the World Health Organization, no. 2, p. 100) and entered into force on 7 April 1948.

Wrzesniewski, A., & Dutton, J. E. (2001). Crafting a job: Revisioning employees as active crafters of their work. *Academy of Management Review*, *26*(2):179–220. https://doi.org/10.5465/amr.2001.4378011

Xu, X., Jiang, L., & Wang, H. J. (2019). How to build your team for innovation? A cross-level mediation model of team personality, team climate for innovation, creativity, and job crafting. *Journal of Occupational and Organizational Psychology*, *92*(4):848–872. https://doi.org/10.1111/joop.12277

Yang, J. (2025). The application of a behavior-based safety program at power plant sites: A pre-post study. *International Journal of Clinical Case Reports and Reviews*. https://doi.org/10.31579/2690-4861/664

Yang, J., & Kwon, Y. (2022). The application of a behavior-based safety program at power plant sites: A pre-post study. Preprints. https://doi.org/10.20944/preprints202202.0231.v1

Yari, S., Naseri, M. H., Akbari, H., Shahsavari, S., & Akbari, H. (2019). Interaction of safety climate and safety culture: A model for cancer treatment centers. *Asian Pacific Journal of Cancer Prevention*, *20*(3), 961–969. https://doi.org/10.31557/APJCP.2019.20.3.961

Zadow, A. J., Dollard, M. F., Dormann, C., & Landsbergis, P. (2021). Predicting new major depression symptoms from long working hours, psychosocial safety climate and work engagement: A population-based cohort study. *BMJ Open*, *11*(6). https://doi.org/10.1136/bmjopen-2020-044133

Zahiri Harsini, A., Bohle, P., Matthews, L. R., Ghofranipour, F., Sanaeinasab, H., Shokravi, F. A., & Prasad, K. (2021). Evaluating the consistency between conceptual frameworks and factors influencing the safe behavior of Iranian workers in the petrochemical industry: Mixed methods study. *JMIR Public Health and Surveillance*, 7(5), e22851. https://doi.org/10.2196/22851

Zakaria, J., Hassan, C. R. C., Hamid, M. D., & Sukadarin, E. H. (2023). The effectiveness of behavior-based safety observation program (BSOP) in the chemical manufacturing industry. *Process Safety Progress*, *43*. https://doi.org/10.1002/prs.12533

Zhang, W., Zeng, X., Liang, H., Xue, Y., & Cao, X. (2023). Understanding how organizational culture affects innovation performance: A management context perspective. *Innovation Management and Organizational Performance for Sustainable Future*, *15*(8). https://doi.org/10.3390/su15086644

Znajmiecka, M., Roszko-Wójtowicz, E., & Stasiła-Sieradzka, M. (2022). In search of differences in the perception of safety climate by employees of an international manufacturing company. *International Journal of Environmental Research and Public Health*, *19*(22), Article 14980. https://doi.org/10.3390/ijerph192214980

Znajmiecka-Sikora, M. (2014). Ocena kultury bezpieczeństwa w sektorze MISP na podstawie badań przedsiębiorstw z regionu łódzkiego [Assessment of safety culture in the SME sector based on a study of enterprises from the Łódź region]. In A. Jużwicka, K. Szymańska, & A. Walecka (Eds.), *Nowe spojrzenie na kulturę organizacyjną* [A new perspective on organizational culture] (pp. 53–62). Wydawnictwo Politechniki Łódzkiej.

Znajmiecka-Sikora, M. (2019). Właściwości psychometryczne skróconej wersji Kwestionariusza Klimatu Bezpieczeństwa-50 (KKB-50) [Psychometric properties of the shortened Safety Climate-50 Questionnaire]. *Zeszyty Naukowe Politechniki Łódzkiej. Organizacja i Zarządzanie*, *72*, 73–94. https://doi.org/10.34658/oiz.2019.72

Znajmiecka-Sikora, M., & Boczkowska, K. (2016). Analysis of safety culture on the example of selected Polish production enterprises. In P. Arezes, J. S. Baptista, M. P. Barroso, P. Carneiro, P. Cordeiro, N. Costa, R. B. Melo, A. S. Miguel, & G. Perestrelo (Eds.), *Occupational safety and hygiene IV* (pp. 349–352). CRC Press.

Znajmiecka-Sikora, M., & Boczkowska, K. (2017). Behaviour-based safety (BBS): Program modyfikacji zachowań niebezpiecznych w przedsiębiorstwie [Behaviour-based safety (BBS): A dangerous behavior modification program in an enterprise]. *Bezpieczeństwo Pracy – Nauka i Praktyka*, *3*. https://yadda.icm.edu.pl/baztech/element/bwmeta1.element.baztech-c22ea7ca-cf8d-49f9-8d68-01db2c0f9423

Zohar, D. (1980). Safety climate in industrial organizations: Theoretical and applied implications. *Journal of Applied Psychology*, *65*(1), 96–102.

Zohar, D. (2010). Thirty years of safety1 climate research: Reflections and future directions. *Accident Analysis & Prevention*, *42*(5). https://doi.org/10.1016/j.aap.2009.12.019

Zohar, D., Luria, G. (2005). A multilevel model of safety climate: Cross-level relationships between organization and group-level climates. *Journal of Applied Psychology*, *90*(4), 616–628. https://doi.org/10.1037/0021-9010.90.4.616

Zohar, D., & Polachek, T. (2014). Discourse-based intervention for modifying supervisory communication as leverage for safety climate and performance improvement: A randomized field study. *Journal of Applied Psychology*, *99*(1), 113–124. https://doi.org/10.1037/a0034096

Index

Pages in **bold** refer to tables.

Q

R

S

T

W

For Product Safety Concerns and Information please contact our
EU representative GPSR@taylorandfrancis.com Taylor & Francis
Verlag GmbH, Kaufingerstraße 24, 80331 München, Germany